Bistatic SAR Clutter Suppression

Zhongyu Li · Junjie Wu · Jianyu Yang · Zhutian Liu

Bistatic SAR Clutter Suppression

Theory, Method, and Experiment

 Springer

Zhongyu Li
University of Electronic Science
and Technology
Chengdu, China

Junjie Wu
University of Electronic Science
and Technology
Chengdu, China

Jianyu Yang
University of Electronic Science
and Technology
Chengdu, China

Zhutian Liu
University of Electronic Science
and Technology
Chengdu, China

ISBN 978-981-19-0161-4   ISBN 978-981-19-0159-1 (eBook)
https://doi.org/10.1007/978-981-19-0159-1

This Springer imprint is published by the registered company Springer Nature Singapore Pte Ltd.
The registered company address is: 152 Beach Road, #21-01/04 Gateway East, Singapore 189721,
Singapore

Preface

With the development of synthetic aperture radar (SAR) technology, bistatic SAR has been proposed. Since the radar receiver and transmitter are placed on different platforms, bistatic SAR has advantages of concealment, stronger antijamming ability, and more abundant characteristics compared with monostatic SAR. Under the increasing requirements, bistatic SAR is also applied in ground/maritime moving target detection, because moving target information is important for surveillance systems. Bistatic SAR clutter suppression is the prerequisite and key technique for realizing the detection of moving targets on the ground/sea surface. It is also the frontier and difficult issue of international research in the field of radar.

This book reports the latest results in the study of clutter suppression and signal processing techniques and focuses on the analysis of non-stationary characteristics of bistatic SAR clutter, DPCA-based clutter suppression method, optimization-based clutter suppression method, sparse-recovery-based clutter suppression method, experimental verification, and many other aspects, i.e., including the research results of realization technology and experimental verification. Researchers, engineers, and graduate students in Radar Signal Processing can benefit from this book, who wish to learn the core theories, methods, and applications of bistatic SAR technologies.

In Chap. 1, this book presents an overview of bistatic SAR development and technologies. Chapter 2 introduces the basic theory of clutter suppression. Chapter 3 gives signal models of bistatic SAR, including a signal model in the time domain and a clutter model in the space-time domain. Chapter 4 introduces DPCA-based methods for BiSAR clutter suppression. Chapter 5 introduces the optimization-based method for BiSAR clutter suppression. Chapter 6 introduces the spare-recovery-based method for BiSAR clutter suppression. Finally, Chapter 7 introduces the experimental technique of University of Electronic Science and Technology of China (UESTC) in BiSAR-MTD. In October 2020, the first airborne BiSAR-MTD experiment in the world has been successfully conducted by UESTC. The experiment situation is introduced in this chapter as well and the experimental data is also applied to validate the effectiveness of the method proposed in this book.

Chengdu, China
January 2022

Zhongyu Li

Acknowledgements

The authors would like to express their sincere thanks to Dr. Qing Yang, Dr. Junao Li, Mr. Hongda Ye, Mr. Xiaodong Zhang, Dr. Xingye Qiu, Mr. Fei Yu, Ms. Yahui Wang, Dr. Zhichao Sun, Dr. Hongyang An, Prof. Yulin Huang, Prof. Haiguang Yang, and Prof. Yin Zhang.

The author thanks the editors of this series and the Springer team for their valuable guidance and assistance.

Contents

Chapter 1
Introduction

Abstract This chapter introduces bistatic SAR development and technologies, and summarizes existing clutter suppression methods, including single-channel clutter suppression methods and multi-channel clutter suppression methods. Additionally, it also describes existing SAR moving target detection systems. The structure of this book is also shown at the end of this chapter.

Keywords Bistatic SAR · Clutter suppression · Single-channel methods · Multi-channel methods · SAR moving target detection systems

1.1 Bistatic SAR Technologies

Different with monostatic SAR, the transmitter and the receiver of bistatic SAR (BiSAR) are mounted on two platforms to provide radar images by the principle of cooperative synthetic aperture, which is one branch of SAR technology [1, 2]. Based on the observing direction of the receiver while imaging, BiSAR can be categorized into side-looking SAR, squint-looking SAR, downward-looking SAR, backward-looking SAR and forward-looking SAR, each of which has a different aim in practice [3]. Generally, bistatic side-looking SAR and bistatic squint-looking SAR are suitable for scouting. Bistatic downward-looking SAR and bistatic backward-looking SAR perform well in assessment applications. And, bistatic forward-looking SAR is applicable to attack enemies. In the applications mentioned above, although the requirements of timeliness and mechanical time consumption in BiSAR imaging are various, all of them are practically valuable and promising.

The earliest research of BiSAR can be dated back to 1970s. An American company, called Xonics, in 1977, pointed out that after separating the transmitter and the receiver, side-looking imaging in bistatic SAR could be achieved by a certain geometry configuration [4, 5]. In 1979, supported by the Defense Advanced Research Projects Agency (DARPA) and the United States Air Force (USAF), Xonics and Goodyear Aerospace company took the lead in carrying out the theoretical research and system development of BiSAR, and eventually obtained well-focused bistatic side-looking SAR image in 1983 [4]. In twenty-first century, with the great development of the time–frequency synchronization system as well as the continuous technical innovations of navigation, orientation and communication, BiSAR has become increasingly attractive for the countries all over the world, due to its unique advantages in modern diversified remote sensing and detection fields. Moreover, some famous international conferences, such as IEEE Radar conference, IEEE IGARSS, EUSAR, APSAR and so on, have set up the topic of BiSAR technology. A series of important plenary lectures were held and significant academic achievements were obtained in these conferences [6–9].

In the beginning of twenty-first century, at IEEE IGARSS international conference held in France, an idea was stated by Deutsches Zentrum für Luft-und Raumfahrt (DLR): BiSAR could provide images of the forward-looking terrain of the receiver[10]. From then on, bistatic forward-looking SAR (BFSAR) technology has been extremely popular around the world and been thriving and prosperous. Only one year later, Sandia National Laboratories (SNL) decided to list BFSAR as one of its most important research directions, since BFSAR technology would make great progress on avionics reconnaissance [11]. In 2005, Fraunhofer Institute for High Frequency Physics and Radar Techniques (FHR) also pointed out that one essential strength of BiSAR, compared with monostatic SAR, was its ability to image the forward-looking terrain of the transmitter or the receiver [12].

Two X-band SAR satellites were launched by Deutsches Zentrum fur Luft-und Raumfahrt (DLR) between 2007 and 2010, constructing the spaceborne BiSAR system, named TanDEM-X. In December 2007, DLR utilized TerraSAR-X satellite and F-SAR as the transmitter and the airborne receiver, respectively, and carried out a spaceborne-airborne side-looking BiSAR imaging experiment. From July 2008 to early 2009, Fraunhofer Institute for High Frequency Physics and Radar Technology (FHR) carried out two spaceborne BiSAR imaging experiments, where TerraSAR-X satellite was the transmitter and PAMIR airborne radar system was the receiver, via different working mode of the receiving antenna and signal bandwidth. To verify the feasibility of BFSAR imaging, in November 2009, FHR conducted a spaceborne-airborne backward-looking imaging experiment to simulate forward-looking imaging [13–15], using the transmitter of TerraSAR-X satellite and the receiver of PAMIR, as shown in Fig. 1.1.

During 2012–2013, FHR carried out a further BiSAR experiment for forward-looking radar imaging, where the transmitter was a ground moving automobile and the receiver was an aircraft [16, 17]. The BiSAR configuration and image result of the airport runway are shown in Fig. 1.2.

As for the BiSAR research in China, from 2003, several famous universities, such as University of Electronic Science and Technology of China (UESTC), National University of Defense Technology (NUDT), Xidian University (XDU) and Beijing Institute of Technology (BIT), have made great promotions and contributions. In 2004, UESTC started the research of imaging theory and implementation of airborne BiSAR, and then manufactured a prototype of the experimental system. In 2007, UESTC carried out BiSAR experiment and obtained the first image provided by bistatic side-looking SAR in China [18–20], as depicted in Fig. 1.3. Until

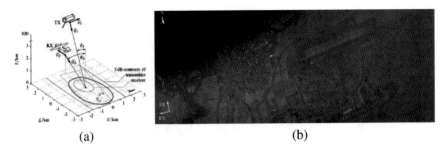

(a)   (b)

Fig. 1.1 Spaceborne-airborne BiSAR experiment for backward-looking radar imaging carried by FHR [15]. **a** Spaceborne-airborne BiSAR configuration; **b** Imaging result

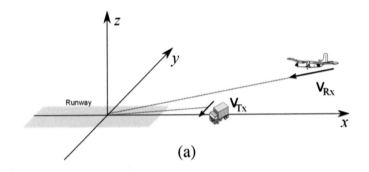

(a)

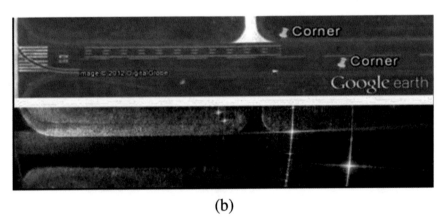

(b)

Fig. 1.2 The BiSAR experiment with a car and an aircraft carried by FHR [16]. **a** Airborne-carborne BiSAR configuration; **b** Imaging result

now, UESTC has made tremendous progress on BiSAR system structures, resolution theories, imaging algorithms, configuration design methods and motion error compensation techniques.

Fig. 1.3 Imaging result of airborne bistatic side-looking SAR obtained by UESTC in 2007 [3]

In 2012, UESTC organized the first BFSAR experiment in the world in Shanxi province and successfully obtained the first airborne BFSAR image [21]. In BFSAR experiment, the transmitter which was squint-side-looking, and the receiver that was forward-looking, were mounted on two Yun-5 transport planes. This experiment had effectively verified BiSAR's ability of high-resolution imaging of the forward-looking terrain. During the period between 2013 and 2020, UESTC undertook several BFSAR experiments in Zhengzhou, Chengdu and Yinchuan, further improving the resolution and quality of BFSAR images. Figure 1.4 shows the world's first airborne BFSAR image obtained by UESTC. Figure 1.5 shows the imaging result in another airborne BFSAR experiment, which was obtained in Yinchuan in 2020.

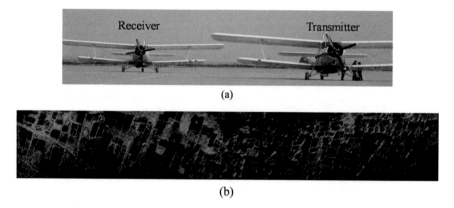

Fig. 1.4 World's first airborne BFSAR image obtained by UESTC in 2012 [21]. **a** Platforms in world's first BFSAR experiment; **b** Imaging result

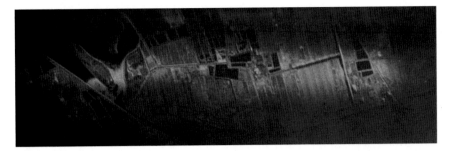

Fig. 1.5 The imaging result of airborne BFSAR experiment in Yinchuan in 2020

1.2 Clutter Suppression Methods

Based on the number of receiving channels, clutter suppression method can be divided into single-channel methods and multi-channel methods, which will be analyzed and summarized in the following.

1.2.1 Single-channel Methods

Single-channel clutter suppression method utilizes only one receiving channel for separation of moving target and clutter via their Doppler characteristic differences. In 1971, an American scientist called R. K. Raney was the first to research the idea of detecting and imaging ground moving target by airborne SAR system [22]. He analyzed the phase and frequency characteristics of moving target, and proposed several detecting methods, beginning a research era of SAR clutter suppression and target detection. Single-channel methods can be classified into following types:

(1) Methods based on Doppler filtering: Since the radial velocity of moving target leads to an additional Doppler centroid, clutter suppression can be realized via filtering by using the Doppler centroid difference between clutter and moving target. The methods in literature [23, 24] are the typical examples of the single-channel clutter suppression method based on Doppler filtering.

(2) Methods based on Doppler frequency rate (DFR) filtering: Since the acceleration and the along-track velocity of moving target will make changes on its DFR, clutter suppression can be realized by using DFR difference between clutter and moving target in this type of methods. Typical examples of the single-channel clutter suppression method based on DFR filtering can refer to [25, 26].

(3) Method based on multi-view processing: By segmenting the signal spectrum and incoherently subtracting, clutter can be suppressed and moving target can

be detected via using this type of methods. Typical examples of the single-channel clutter suppression method based on multi-view processing can refer to [27, 28].

(4)   Methods based on time–frequency analysis: This type of methods is actually the combination of type 1 and type 2. It jointly uses signals' Doppler centroid and DFR in clutter suppression and target detection, which mainly contains Wigner-Ville distribution (WVD) [29, 30], WVD-Hough transform [31], Lv's distribution [32], fractional Fourier transform [33, 34] and so on.

(5)   Methods based on variation detection: For the imaging results of the same place at different time, clutter signal is unchanged. However, because of target's velocity, image signal of moving target varies with azimuth time. Thus, according to the variation of target's image signal, clutter suppression and target detection can be achieved by this type of method [35].

Although single-channel clutter suppression method has advantages of simple hardware system and low computational complexity, it is generally difficult to suppress clutter and detect moving target with a low velocity, as they are mixed with mainlobe clutter. Additionally, in a BiSAR system, both moving target and clutter have spectrum expansion. As a concequense, their spectrums in Doppler domain might be aliased and hardly be distinguished. Thus, via single-channel methods based on Doppler difference, target signal may be suppressed as well as clutter, which causes the performance degradation of the following target detection.

1.2.2   Multi-channel Methods

Considering the restrictions of single-channel clutter suppression methods mentioned above, multi-channel clutter suppression methods were born. Multi-channel method can use the spatial degree of freedom (DOF) to obtain spatial frequency information and extend different dimensions between moving target and clutter. Thus, clutter echo can be effectively suppressed in two-dimensional processing domain, and moving target with low velocity can be detected. The common multi-channel clutter suppression methods are demonstrated in the following.

(1)   Displaced Phase Center Antenna (DPCA) technology.

DPCA technology has been widely used since 1980s [36–39]. Until now, DPCA is one of the most popular clutter suppression methods.

DPCA usually utilizes two along-track channels to receive ground echo data. By phase compensation and calibration, signals received by two channels can be equivalent to the echo data at different time but in the same place. In monostatic SAR-DPCA, channel 1 is self-transmitting and self-receiving, whose phase center is located at its spatial position, whereas channel 2 only receives echoes, whose equivalent phase center is approximately at the midpoint of channel 1 and channel 2. After compensation and calibration, the phase centers of two channels are coincided with each other. The background echo in two receiving channels, i.e., clutter, is the same, while the echo of moving target will have additional difference between two channels due to its movement. Thus, via subtracting the echo data in channel 1 with that in channel 2, clutter echo can be suppressed and moving target signal can be preserved.

Traditional DPCA method has a strict requirement about the platform velocity, the channel interval and the pulse repetition rate (PRF), namely, DPCA condition. This condition will directly cause the limit of the applications of DPCA. With the development and innovation of technology, a series of modified DPCA methods have been proposed by researchers [40–43], evolving into forms like multiple phase centers-DPCA, frequency domain DPCA and adaptive DPCA. In these methods, not only moving target can be detected, but also their parameters can be precisely estimated.

However, since the transmitter and the receiver in BiSAR configuration are separated, the equivalent phase centers of two receiving channels can hardly be coincided in space, which means DPCA condition is difficult to be satisfied. In addition, the equivalent phase centers in BiSAR vary with space and time, thus the phase error between two channels cannot be centrally compensated. As a consequence, BiSAR clutter cannot be effectively cancelled by the DPCA methods mentioned above. Furthermore, the most important restriction of DPCA is channel consistency.

(2) Space Time Adaptive Processing (STAP) method.

In 1973, the concept of space–time two-dimensional adaptive processing was firstly presented by American scientists Brennan, Mallett and Reed [44, 45]. Different with echo cancellation in DPCA method, which belongs to deterministic signal processing, STAP is an adaptive processing method depending on the training data, and it can achieve the optimal suppression performance according to the statistical characteristic of clutter. In the multi-channel system (where the number of channels is ≥ 2), STAP has a better clutter suppression effect than DPCA, as well as stronger robustness. The basic principle of STAP is to take full advantage of the space–time coupling characteristic of clutter signals received by array radars, and adaptively filter echo data via two-dimensional joint processing in space–time domain [46, 47]. After applying STAP, the output signal-to-clutter-and-noise ratio (SCNR) can be maximized. In short, STAP can effectively suppress clutter and interference, and preserve the energy of moving target as much as possible, which enormously increases target detection ability of airborne radars.

In early stage, STAP method was applied to airborne early warning (AEW) aircraft to suppress clutter and interference. Due to the clutter suppression performance of STAP, it has been well applied to airborne SAR and spaceborne SAR as well. In airborne SAR-STAP, Ender et al. were the first group of people who applied STAP to multi-channel SAR systems [48]. In spaceborne SAR-STAP, Cerutti-Maori et al. verified the clutter suppression performance of the imaging STAP (ISTAP) method via experimental spaceborne SAR data processing [49, 50].

However, problems of large computational complexity and poor estimation accuracy of covariance matrices have always been existing, ever since the appearance of STAP, which severely restrict its engineering implementation.

In order to improve the processing efficiency of STAP, researchers around the world proposed many methods of dimension reduction and rank reduction. In the 1980s, Dr. R. Klemm from Germany created a novel theory about the fixed reduced dimension STAP and proposed the auxiliary channel approach [51], which could reduce DOF of the system. However, the auxiliary channel approach is sensitive to the error. In the 1990s, the academician Bao proposed the method called mDT [52], which was with highly theoretical significance and practical value. The technical report of Ward [53] and the monographs written by Wang in 2000 [54], R. Klemm in 2006 [55] and J. R. Guerci in 2015 [56] have analyzed and concluded various methods of reduced dimension STAP and reduced rank STAP.

For another problem, the estimation accuracy of clutter covariance matrix (CCM) directly determines the performance of STAP. According to the Reed-Mallet-Brennan (RMB) rule [45], STAP requires samples with more than twice DOF independently and identically distributed (IID) to ensure that the output performance corresponds to 3 dB level below the optimum. However, this rule is hard to be satisfied in practice. Due to the separation of the transmitter and the receiver, BiSAR clutter is strong non-stationary [57, 58]. Clutter in different range cells has different space–time characteristics, which leads to poor estimation accruacy of CCM. In consequence, it's difficult to obtain a satisfied result of clutter suppression via STAP processing in BiSAR.

1.3 SAR Moving Target Detection Systems

SAR moving target detection (SAR-MTD) system has been researched and applied for several years. Its foundations and applications are the useful references for the further study about BiSAR-MTD systems. This section will make a brief introduction about several typical airborne/spaceborne MTD radar systems in the world.

1.3.1 American Airborne MTD Radar System

In 1991, U. S. Air Force and U. S. Army developed the most representative airborne MTD radar system, called "Joint Surveillance and Target Attack Radar System" (JSTARS System) [59–61]. Its core component was the airborne SAR/MT multimode passive phased array radar (AN/APY-7). JSTARS was mainly used to monitor ground moving targets, where the three-channel interference algorithm was used to suppress mainlobe clutter and realize target indication [59]. In addition, JSTARS could also estimate target's velocity, providing the command department with overall dynamic information of the battlefield and leading the troops to make attacks precisely.

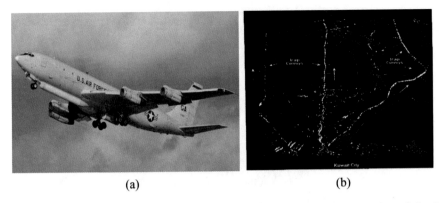

(a)   (b)

Fig. 1.6 Airborne JSTARS and MTD processing result. **a** Airborne JSTARS platform; **b** Iraqi army's withdrawal in 1991 [61]

During the Gulf War in 1991, two E-8A planes carried JSTARS were sent to the battlefield participating the mission "Desert Storm", which was the milestone of MTD application. In this military action, JSTARS played an important role to monitor real-time movements of Iraqi ground maneuver troops, tanks and scud missiles. Figure 1.6a shows the airborne JSTARS platform and Fig. 1.6b is the intelligence of Iraqi army withdrawal during the Gulf War acquired by JSTARS.

Besides, the advanced fighter F-22 was equipped with AN/APG-77 [62, 63], which had multi-channel MTD ability, making this kind of fighters able to indicate ground moving targets. In 2006, in Fort Worth, Texas, American mounted the AN/APG-81 radars, that is, the upgrade of AN/APG-77, on F-35 fighters and carried out flight tests. AN/APG-81 made F-35 fighters have the abilities of high-resolution SAR imaging, multi-target detection and tracking. F22/F-35 multi-mode, multi-mission radar capabilities are shown in Fig. 1.7.

1.3.2 German Airborne/Spaceborne MTD Radar System

AER-II system [64] and PAMIR system invented by Forschungsgesellschaft für Angewandte Naturwissenschaften (FGAN) are the representative Germany airborne MTD radar systems.

AER-II system worked in X-band (center frequency: 10 GHz) and was equipped with four parallel receiving channels with the maximum detection range of 20 km, involving working modes such as single-channel SAR and four-channel MTD. For SAR imaging mode, AER-II system could obtain images of stationary scene with the resolution of 1 m × 1 m, while for four-channel MTD mode, ground clutter could be suppressed via adaptive processing technology. Figure 1.8 shows the AER-II system and MTD processing result.

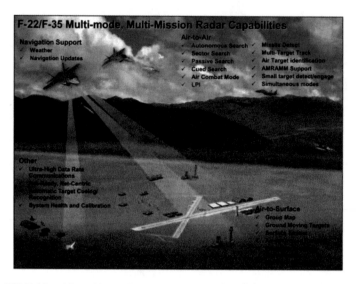

Fig. 1.7 F22/F-35 multi-mode, multi-mission radar capabilities [62]

(a)   **(b)**

Fig. 1.8 AER-II system [64]. **a** Installation of AER-II radar; **b** MTD processing result

In 2002, FGAN developed the upgrade of AER-II system—PAMIR system [64–67], which instantly attracted global attentions. PAMIR system worked in X-band (central frequency: 9.45 GHz) with signal bandwidth of 1820 MHz and detection range of 100 km. It had five parallel receiving channels and was capable of multi-channel SAR imaging, super-resolution SAR imaging, three-dimensional SAR imaging, full-polarization SAR imaging and wide area MTD. In the imaging mode, PAMIR system could realize the resolution of 0.08 m. In MTD processing, PAMIR system used space–time adaptive processing technology to suppress ground clutter and estimate the velocity of moving target (smaller than 1 m/s). PAMIR realized the

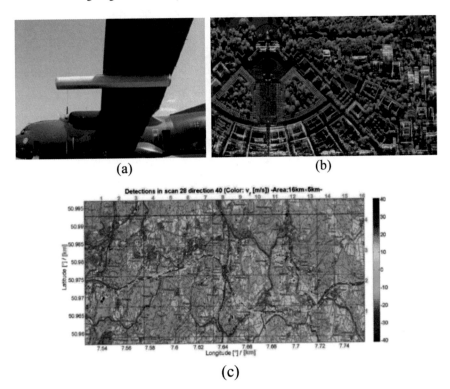

(a)

(b)

(c)

Fig. 1.9 PAMIR system. **a** C-160 transport plane with PAMIR system [65]; **b** Imaging result of PAMIR system [66]; (c) MTD result of PAMIR system [67]

SCAN-MTD mode by beam scanning (azimuth scanning range: $\pm 45°$), to achieve the goal of monitoring moving target in wide area [67]. Figure 1.9 shows PAMIR system and its processing results.

Moreover, DLR successfully launched a TerraSAR-X satellite in 2007 and constructed a spaceborne monostatic MTD system. Its orbit altitude was 514 km and working frequency was X-band, with high-resolution wide-swath imaging mode and MTD mode [68, 69]. TerraSAR-X satellites could obtain SAR images with a resolution of 3 m and a swath width of 30×50 km in strip imaging mode, while SAR could only image with a resolution of 0.21 m and a swath width of 2.5×4 km in the spotlight imaging mode. Under MTD mode, TerrasAR-X combined DPCA with ATI to improve moving target detection performances. Figure 1.10 shows the TerrasAR-X system and its processing results.

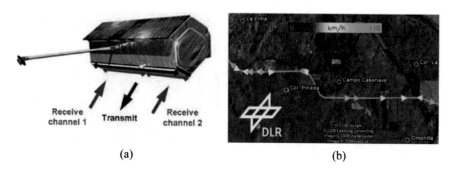

                (a)   (b)

Fig. 1.10 TerraSAR-X system and its MTD processing result. **a** Schematic of TerraSAR-X satellite [68]; **b** Detection of traffic condition [69]

1.3.3 Chinese Airborne BiSAR-MTD Radar System

From September to October in 2020, UESTC has carried out the first airborne BiSAR-MTD experiment in Yinchuan, China, in the world. Figure 1.11 shows two Cessna-208 planes used as the transmitter and the receiver respectively. In this experiment, the transmitter worked in squint-looking mode while the receiver worked in forward-looking mode. Through the airborne experiment, UESTC verified the mechanical

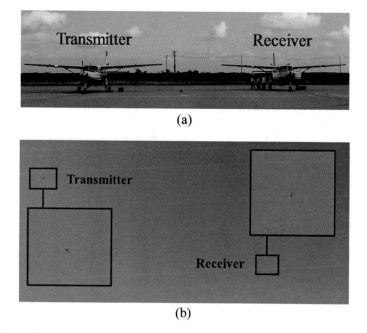

Fig. 1.11 World's first airborne BiSAR-MTD experiment carried by UESTC in 2020. **a** Platforms in the world's first airborne BiSAR-MTD experiment; **b** Flights photo

feasibility of BiSAR-MTD. After processing, BiSAR clutter could be effectively suppressed and moving target could be detected as well as refocused at its real position.

1.4 Structure of This Book

The content of this book is organized as follows.

Chapter 2 introduces the basic theory of clutter suppression, including DPCA fundamentals, STAP fundamentals and their main issues in BiSAR clutter suppression.

Chapter 3 gives signal models of BiSAR, including a signal model in time-domain and a clutter model in space–time domain. Based on the established models, BiSAR echo characteristics are analyzed in this chapter, such as RCM characteristic, Doppler characteristic and clutter non-stationarity.

Chapter 4 introduces DPCA-based methods for BiSAR clutter suppression. Two kinds of DPCA-based methods are shown in this chapter. The multi-pulse DPCA clutter suppression method is constructed in echo domain, and it mainly includes Keystone transform, non-linear chirp scaling processing and multi-pulse canceller construction. Then, image-domain DPCA clutter suppression method is given, including range Doppler-based DPCA (RD-DPCA) and back-projection-based DPCA (BP-DPCA). These two kinds of DPCA methods are not affected by the nonstationary aspect of BiSAR clutter and they do not need to satisfy the strict DPCA condition.

Chapter 5 introduces optimization-based method for BiSAR clutter suppression. The main idea of the proposed method is to directly design and generate a suppression filter in space–time domain, whose space–time frequency response is matched with clutter spectrum. It mainly involves filter design, optimization problem construction, optimal weight solution and matched space–time filtering. Since the generation of the designed filter by this method circumvents CCM estimation, it will not be affected by the non-stationary characteristic of BiSAR clutter as well. This method can be conveniently used to guide the space–time filter design for the effective BiSAR clutter suppression, which is highly desirable in practical applications.

Chapter 6 introduces spare-recovery-based method for BiSAR clutter suppression. This method requires only a few secondary samples, which is not affected by the strong non-stationary characteristic of BiSAR clutter in heterogeneous environments. It mainly involves RCM correction, clutter ridge matched dictionary reconstruction, multiple measured vector (MMV) optimization problem construction and clutter filtering. This method can avoid the performance degradation in clutter suppression caused by the off-grid problem and overcome the strong non-stationary problem of BiSAR clutter in heterogeneous environments.

Chapter 7 introduces the experimental technique of UESTC in BiSAR-MTD. In this chapter, system architecture, experimental scheme, experimental results in BiSAR imaging and clutter suppression are detailed. In October 2020, the first airborne BiSAR-MTD experiment in the world has been successfully conducted by UESTC. The experiment situation is introduced in this chapter as well and the experimental data is also applied to validate the effectiveness of the method proposed in Chaps. 5 and 6.

Chapter 2
Basic Theory for Clutter Suppression

Abstract This chapter introduces the basic theories for clutter suppression, including displaced phase center antenna (DPCA) processing fundamentals and space–time adaptive processing (STAP) fundamentals. This chapter also analyzes the main issues in bistatic SAR clutter suppression, which are the fundamentals of the following chapters.

Keywords Bistatc SAR · Clutter suppression · Displaced phase center antenna · Space–time adaptive processing · STAP

2.1 DPCA Processing Fundamentals

Moving target detection method based on DPCA increases spatial information and utilizes the principle of antenna phase center compensation [36–39], to make the system obtain the same clutter information in different time domain and space domain. Thus, clutter can be suppressed by echo cancellation and moving target information can be retained due to its motion.

Traditional DPCA requires at least two phase centers to cancel clutter signal. Take monostatic SAR with two channels for example, the principle of DPCA is shown in Fig. 2.1.

Assume that radar system is working in side-looking mode. Channel 1 and channel 2 are placed along the moving direction of the platform. Channel 1 is self-transmitting and self-receiving, whereas channel 2 only receives echo signal. The channel space is d, the platform velocity is V and the pulse repetition frequency is f_r.

According to the principle of DPCA in Fig. 2.1, when the first echo arrives, the phase center of channel 1 is located at its spatial position, i.e., O_1. The equivalent phase center of channel 2 is approximately at the midpoint of two channels, i.e., O. Assume that radar platform moves with the distance of $d/2$, when the mth pulse is received. At this time, the phase center of channel 1 has moved to O_1' and the phase center of channel 2 has moved to O'. It can be seen that the phase center of channel 1 at the first pulse receiving time and the phase center of channel 2 at the mth pulse receiving time are coincided in space. Thus, the background echo in two receiving channels, i.e., clutter, is the same, while the echo of moving target

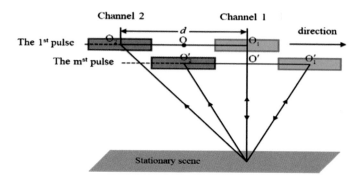

Fig. 2.1 The principle of displaced phase center antenna processing

will have additional difference between two channels due to its movement. Based on this characteristic, with time calibration and phase compensation, clutter echo can be effectively suppressed via subtracting the echo data in channel 1 with that in channel 2. And moving target signal can be retained due to its additional difference between channels. Clutter suppression process mentioned above needs the multi-channel system strictly meet the DPCA condition [70]:

$$d = 2k \cdot V T_r = \frac{2k \cdot V}{f_r} \tag{2.1}$$

where k is a positive integer and T_r is the pulse repetition interval. In practice, since the DPCA condition is hard to be satisfied, the phase centers of receiving channels cannot be coincided, which directly limits the performance and application of DPCA.

In the following, the traditional DPCA method is described in detail. The instantaneous slant ranges between receiving channels and moving target can be expressed as [71]

$$R_1(\eta) = \sqrt{(x_0 + v_a\eta - V\eta)^2 + (y_0 - v_y\eta)^2 + H^2}$$
$$\approx R_B - v_r(\eta - \eta_c) + \frac{(v_a - V)^2(\eta - \eta_c)^2}{2R_B} \tag{2.2}$$

$$R_2(\eta) = \sqrt{(x_0 + v_a\eta - V\eta + d)^2 + (y_0 - v_y\eta)^2 + H^2}$$
$$\approx R_B - v_r(\eta - \eta_c) + \frac{[(v_a - V)(\eta - \eta_c) + d]^2}{2R_B} \tag{2.3}$$

After demodulating the moving target echo signals received from the two channels, they can be expressed as

$$s_1(\tau, \eta) = a_r(\tau - \tau_1)a_a(\eta)\exp\left[j\pi\gamma(\tau - \tau_1)^2\right]\exp\left[-j\frac{4\pi}{\lambda}R_1(\eta)\right] \qquad (2.4)$$

$$s_2(\tau, \eta) = a_r(\tau - \tau_2)a_a(\eta)\exp\left[j\pi\gamma(\tau - \tau_2)^2\right]\exp\left[-j\frac{2\pi}{\lambda}(R_1(\eta) + R_2(\eta))\right] \qquad (2.5)$$

where τ_1 and τ_2 are the echo delay of channel 1 and 2 respectively, which can be expressed as

$$\tau_1 = \frac{2R_1(\eta)}{c}, \tau_2 = \frac{R_1(\eta) + R_2(\eta)}{c} \qquad (2.6)$$

After range pulse compression, the echo signals are expressed as

$$s_1(\tau, \eta) = \sigma_s T_p \text{sinc}(B(\tau - \tau_1))a_a(\eta)\exp\left[-j\frac{4\pi}{\lambda}R_1(\eta)\right] \qquad (2.7)$$

$$s_2(\tau, \eta) = \sigma_s T_p \text{sinc}(B(\tau - \tau_2))a_a(\eta)\exp\left[-j\frac{2\pi}{\lambda}(R_1(\eta) + R_2(\eta))\right] \qquad (2.8)$$

The instantaneous slant range is substituted by the above formula, as shown below

$$s_1(\tau, \eta) = G_1(\tau, \eta)\exp\left(-j\frac{4\pi}{\lambda}R_B\right)\exp\left(j\frac{2\pi}{\lambda}\frac{v_r^2 R_B}{(V - v_a)^2}\right)$$
$$\times \exp\left[j\pi\left(-\frac{2(V - v_a)^2}{\lambda R_B}\right)\left(\eta - \eta_c - \frac{v_r R_B}{(V - v_a)}\right)^2\right] \qquad (2.9)$$

where

$$G_1(\tau, \eta) = \sigma_s T_p \text{sinc}(B(\tau - \tau_1))a_a(\eta)$$
$$G_2(\tau, \eta) = \sigma_s T_p \text{sinc}(B(\tau - \tau_2))a_a(\eta) \qquad (2.10)$$

Considering that the function sinc(·) changes the least at the maximum position, and $\tau_1 \approx \tau_2$, $G_1(\tau, \eta) \approx G_2(\tau, \eta)$ can be obtained. In order to align the phase centers of the two channels, the echo signal of channel 1 is phase compensated, and the expression of the phase compensation function is set as

$$C_1 = \exp\left(-j\frac{\pi d^2}{2\lambda R_B}\right) \qquad (2.11)$$

Next, the channel 2 echo signal is processed with slow time delay, and the delay time is expressed as

$$T = \frac{d}{2(V - v_a)} \approx m/PRF \tag{2.12}$$

After phase compensation and time calibration, DPCA cancellation is performed on the range-compressed echo data, and the following formula can be obtained

$$s_{12}(\tau, \eta) = s_1(\tau, \eta) \times C_1 - s_2(\tau, \eta + T)$$

$$= G(\tau, \eta) \exp\left(-j\frac{4\pi}{\lambda} R_B\right) \exp\left(j\frac{2\pi}{\lambda} \frac{v_r^2 R_B}{(V - v_a)^2}\right)$$

$$\times \exp\left[j\pi\left(-\frac{2(V - v_a)^2}{\lambda R_B}\right)\left(\eta - \eta_c - \frac{v_r R_B}{(V - v_a)}\right)^2\right]$$

$$\times \exp\left(-j\frac{\pi d^2}{2\lambda R_B}\right)\left[1 - \exp\left(j\frac{2\pi}{\lambda} \frac{v_r d}{V - v_a}\right)\right] \tag{2.13}$$

The signal amplitude information obtained from the modulus value of the above formula after cancellation is

$$|s_{12}(\tau, \eta)| = \sigma_s T_p a_a(\eta) |\mathrm{sinc}(B(\tau - \tau_1))| \cdot \left|1 - \exp\left(j\frac{2\pi}{\lambda} \frac{v_r d}{V - v_a}\right)\right|$$

$$= \sigma_s T_p a_a(\eta) |\mathrm{sinc}(B(\tau - \tau_1))| \cdot \left|2\sin\left(\frac{\pi}{\lambda} \frac{v_r d}{V - v_a}\right)\right| \tag{2.14}$$

It can be seen that for the stationary target ($v_r = 0$), the sinc function is zero, so the stationary target is eliminated; For the moving target ($v_r \neq 0$), the sinc function is non-zero so the amplitude information is retained. Thus, after DPCA processing, stationary clutter can be suppressed and moving target can be detected. The sinc function in (2.14) determines the amplitude of moving target after clutter cancellation, which directly affects the signal to clutter and nosie ratio (SCNR) after DPCA processing.

2.2   Space–Time Processing Fundamentals

DPCA method does not sufficiently utilize the multi-channel SAR echo information. Since it has strict requirements on channel consistency and less satisfactory effect of clutter suppression, the multi-channel space–time adaptive processing method has been introduced.

The basic theory of multi-channel space–time adaptive processing is to process echo in space–time domain to obtain the maximum output SCNR under a certain optimal criterion, and finally realize moving target detection. Let multi-channel echo be $Z = [Z_1, Z_2, ..., Z_N]^T$. According to Bernoulli hypothesis testing theorem, a binary hypothesis can be obtained:

$$\begin{cases} H_1 : Z = S + C + N \\ H_2 : Z = C + N \end{cases} \qquad (2.15)$$

where H_1 represents the existence of moving target, H_2 represents the non-existence of moving target, and S, C, N are the echo of moving target, clutter and noise respectively. Assume noise is complex Gaussian signals with zero average value, and the variance of noise is σ_n^2. Noise and clutter components are statistically independent. let R_c and R_n be the covariance matrices of clutter and noise, respectively. The space of noise is mutually unrelated, so $R_n = \sigma_n^2 I_{(N \times N)}$, where $I_{(N \times N)}$ is an $N \times N$ unit matrix. The covariance matrix is:

$$R = E\left[Z_{H0} Z_{H0}^H\right] = R_c + R_n = R_c + \sigma_n^2 I_{N \times N} \qquad (2.16)$$

where $(\cdot)^H$ is the conjugate transposed operation. Because of some unknown factors from system inside and outside surroundings, as well as random errors, in practice, several range cells with independent and identical distribution are selected as training samples for interference covariance estimation under hypothesis H_0, which is called sample matrix inversion (SMI) algorithm. The output of the filter after adaptive processing is:

$$Z_{out} = W^H Z \qquad (2.17)$$

where W is a weight vector. The following picture shows the schematic of space–time adaptive filter (Fig. 2.2).

This adaptive processing structure has N receiving channels, and each channel has K delay units. $\{W_{nk} | n = 1, 2, \ldots, N; k = 1, 2, \ldots, K\}$ is considered as the weight vector of the adaptive filter.

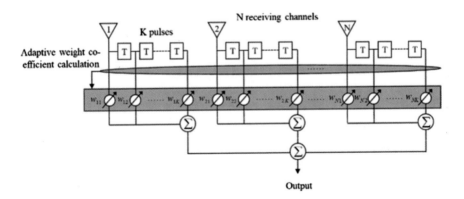

Fig. 2.2 Schematic of space–time adaptive processor

$$W = [w_{11}, w_{12}, \ldots, w_{1K}, w_{21}, \ldots, w_{2K}, \ldots, w_{N1}, w_{N2}, \ldots, w_{NK}]^T \in NK \times 1 \tag{2.18}$$

The filter weight can be obtained by solving the following optimization problem:

$$\begin{cases} \min \ W^H R W \\ s.t. \ W^H S = 1 \end{cases} \tag{2.19}$$

where $R = E\{XX^H\} \in NK \times NK$ is the covariance matrix of clutter and noise. Vector $S = S_s \otimes S_t \in NK \times 1$ is the space–time steering vector. Space steering vector S_s and time steering vector S_t can be expressed as:

$$S_s = \left[1, e^{j\omega_s}, e^{j2\omega_s}, \ldots, e^{j(N-1)\omega_s}\right]^T \tag{2.20}$$

$$S_t = \left[1, e^{j\omega_t}, e^{j2\omega_t}, \ldots, e^{j(K-1)\omega_t}\right]^T \tag{2.21}$$

where ω_s is normalized special frequency and ω_t is normalized time frequency.

By solving the mathematic optimization problem in Eq. (2.19) via Lagrange multiplier method, theoretical optimal weight vector W_{opt} of space–time two-dimensional filter is:

$$W_{opt} = \mu R_c^{-1} S \tag{2.22}$$

where $\mu = 1/(S^H R_c^{-1} S)$ is a constant factor. From Eq. (2.22), it can be seen that the space–time optimal weight vector is depended on the inverse matrix R_c^{-1} and the target steering vector S.

The performances of clutter suppression via STAP method can be analyzed by the output SCNR ($SCNR_{out}$) or the improvement factor (IF), which are defined as:

$$SCNR_{out} = \frac{Pout_{target}}{Pout_{clutter+noise}} \tag{2.23}$$

$$IF = \frac{SCNR_{out}}{SCNR_{in}} = \frac{\frac{Pout_{target}}{Pout_{clutter+noise}}}{\frac{Pin_{target}}{Pin_{clutter+noise}}} \tag{2.24}$$

where $Pout_{clutter+noise}$ and $Pin_{clutter+noise}$ are, respectively, the output and input power of clutter plus noise. $Pout_{target}$ and Pin_{target} are, respectively, the output and input power of target. For the optimal processor, its optimal improvement factor IF_{opt} is the inverse of the minimum variance clutter spectrum, expressed as:

$$IF_{opt} = \frac{S^H R_c^{-1} S}{S^H S} tr(R_c) \tag{2.25}$$

where $tr(\cdot)$ is the inverse of a matrix.

In conclusion, the optimal filter is actually a generalized optimal Wiener filter, whitening the clutter first, and then matching the target signal followed with filtering, to produce a notch at the clutter position and keep a large gain value at the target area. Therefore, STAP technology can effectively suppress clutter and interference, realizing moving target detection.

2.3 Main Issues in Bistatic SAR Clutter Suppression

2.3.1 Problems of DPCA in Bistatic Mode

This section will analyze the problems of DPCA method in BiSAR, based on the range history of two channels.

As shown in Fig. 2.3, $R_T(t_1)$ and $R_T(t_2)$ are instantaneous ranges between the transmitter and the clutter scattering point $P(x, y)$ at time t_1 and t_2 respectively. $R_{R1}(t_1)$ and $R_{R1}(t_2)$ are instantaneous ranges between channel 1 of the receiver and $P(x, y)$ at time t_1 and t_2 respectively. $R_{R2}(t_1)$ and $R_{R2}(t_2)$ are instantaneous ranges between Channel 2 of the receiver and $P(x, y)$ at time t_1 and t_2 respectively. Let t_1 and t_2 satisfy the equation $t_2 = t_1 + mT_r$, so that $d = mVT_r$ where d is the distance between channels and m is a positive integer.

From the basic principle of DPCA method, if the bistatic ranges of the echo in channel 1 at t_1 equals to the one in channel 2 at t_2, then the clutter signals in two channels will have the same information, and thus clutter suppression can be realized by echo subtraction.

At t_1, the bistatic ranges of the echo in channel 1 is:

$$R_1(t_1; x, y) = R_T(t_1; x, y) + R_{R1}(t_1; x, y) \tag{2.26}$$

Fig. 2.3 Range history relationship of two channels in bistatic SAR

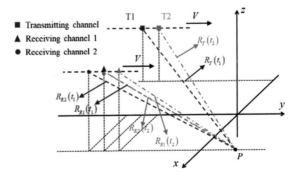

At t_2, the bistatic ranges of the echo in channel 2 is:

$$R_2(t_2; x, y) = R_T(t_2; x, y) + R_{R2}(t_2; x, y) \tag{2.27}$$

After subtracting Eqs. (2.26) and (2.27), the bistatic range difference is:

$$\begin{aligned}
\Delta R &= R_1(t_1; x, y) - R_2(t_2; x, y) \\
&= R_T(t_1; x, y) + R_{R1}(t_1; x, y) - R_T(t_2; x, y) - R_{R2}(t_2; x, y) \\
&= (R_T(t_1; x, y) - R_T(t_2; x, y)) + (R_{R1}(t_1; x, y) - R_{R2}(t_2; x, y))
\end{aligned} \tag{2.28}$$

Then, the following relation can be gained:

$$t_2 = t_1 + mT_r = t_1 + \frac{d}{V} \tag{2.29}$$

which means that during the period between t_1 and t_2, the receiver moves d alone the track. So, the physical phase center of channel 2 at t_2 is at the same position as channel 1 at t_1. The clutter scattering point $P(x, y)$ satisfies:

$$R_{R1}(t_1; x, y) = R_{R2}\left(t_1 + \frac{d}{V}; x, y\right) = R_{R2}(t_2; x, y) \tag{2.30}$$

However, because of the movement of the transmitter, its range relationship relative to $P(x, y)$ will vary with azimuth and time. Thus, contribution about range history of the transmitter will satisfy:

$$R_T(t_1; x, y) \neq R_T(t_2; x, y) \tag{2.31}$$

As a consequence, the bistatic range difference can be simplified as:

$$\Delta R = R_T(t_1; x, y) - R_T(t_2; x, y) = R_T(t_1; x, y) - R_T\left(t_1 + \frac{d}{V}; x, y\right) \tag{2.32}$$

From the bistatic range difference in Eq. (2.32), it can be observed that, due to the separation of the receiver and the transmitter in bistatic SAR, although the receiver satisfies DPCA condition, clutter at the same scene of two receiving channels is still with different range history at different time, and the defference is various with time and space. Therefore, bistatic SAR system is difficult to acquire the same echo of a stationary background in different time domains. Clutter between channels is also different due to the movement of the separated transmitter, causing serious performance loss of DPCA method.

Figure 2.4 shows range history difference of received echo between two receiving channels in bistatic SAR. Due to the separation of the transmitter and the receiver, its range history difference of clutter in different channels can be up to 0.5 m, with corresponding azimuth phase of $\Delta\varphi = 2\pi \cdot \Delta R / \lambda \approx 31.9\pi$, far larger than 0.25π, whose influence cannot be ignored. Figure 2.5 illustrates the DPCA simulation result

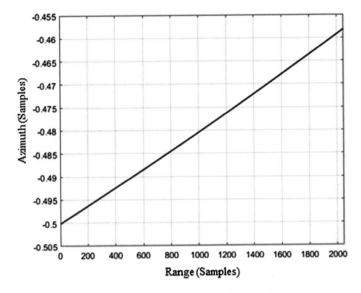

Fig. 2.4 Range history difference between two channels of bistatic SAR

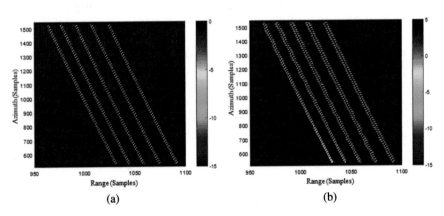

Fig. 2.5 DPCA Simulation in bistatic SAR. **a** After range compression; **b** After DPCA processing

of point targets in bistatic SAR, where Fig. 2.5a is the result after range compression and Fig. 2.5b is the DPCA processing result. It is obvious that the signals of stationary point targets are hard to be suppressed by echo subtraction in bistatic SAR configuration. Thus, moving target cannot be effectively distinguished from clutter, which is consistent with the analyzed conclusion.

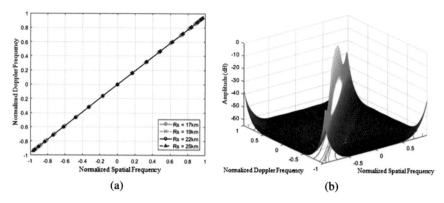

Fig. 2.6 Monostatic SAR clutter characteristic simulation. **a** Clutter ridge; **b** Clutter spectrum distribution

2.3.2 Problems of Traditional STAP in Bistatic Mode

Figures 2.6 and 2.7 respectively show the clutter characteristic simulation results of monostatic SAR and bistatic SAR. The simulation parameters are given in Table 2.1, where the antenna arrays of monostatic SAR and bistatic SAR are installed along trajectory direction.

In the monostatic case, the clutter ridge of different range cells are coincided, as shown in Fig. 2.6. The Doppler frequency of clutter is irrelevant to range, and there is no range correlation. In space–time domain, the distribution of clutter spectrum matches the simulated clutter ridge, which is knife ridge distribution and doesn't change with range cells. Thus, clutter doesn't have non-stationarity in the monostatic case, and it can be well suppressed by the traditional clutter suppression methods.

From the simulation results in Fig. 2.7, a close distance, the clutter ridges are no longer linear distributed. Clutter spectrum ridges in different range cells have different shapes, with a certain degree of position offset. Clutter in different range cells has serious dispersion, and it has strong non-stationarity. At a far distance, the clutter spectrum ridges of different range cells tend to be stable and overlap with each other, and the dispersion and non-stationarity decrease are slow down with the increase of distance. For clutter spectrum of bistatic SAR, it will be significantly broadened with the close distance, which will seriously effect the subsequent moving target detection performance.

Since clutter received by bistatic SAR is non-stationary, clutter in different bistatic range cell has different space–time distribution, thus the samples required in STAP method cannot meet the needs of independent and identical distribution. In practice, when the covariance matrix R_c is estimated by samples in near cells, unbiased estimation cannot be achieved, leading to an error between the estimated result $\hat{R}_c$ and the truth value R_c. Therefore, in bistatic SAR, it is hard to obtain the optimal weight vector of the adaptive filter in Eq. (2.22), which causes degradation of the processing performance of STAP. On the other hand, according to the RMB criterion,

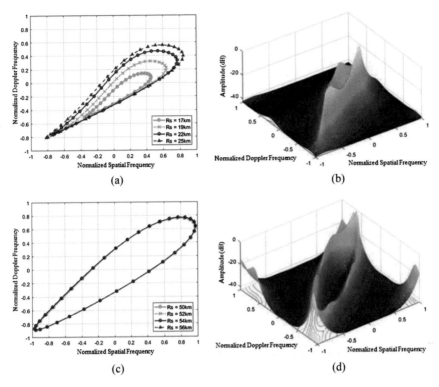

Fig. 2.7 Bistatic SAR clutter characteristic simulation. **a** Clutter spectrum ridge at close distance; **b** Clutter spectrum distribution at close distance; **c** Clutter spectrum ridge at far distance; **d** Clutter spectrum distribution at far distance

Table 2.1 Simulation parameters

| System types | Parameters | | | |
	Transmitter position (km)	Receiver position (km)	Transmitter velocity (m/s)	Receiver velocity (m/s)
Monostatic SAR	$(0, -2, 6)$	$(0, -2, 6)$	$(0, 80, 0)$	$(0, 80, 0)$
Bistatic SAR	$(10, -1, 6)$	$(0, -2, 6)$	$(80, 80, 0)$	$(0, 80, 0)$

to guarantee the actual processing performance less than the optimal one by no more than 3 dB, the sample number for the covariance matrix estimation should be at least twice of the systematic degree of freedom (NK). With the nonstationary bistatic SAR clutter, it is difficult to acquire adequate training samples, affecting the clutter suppression effect of bistatic SAR.

Figure 2.8 shows how the STAP processing performance is affected by the nonstationarity of bistatic SAR clutter. In the simulation, the system DOF is $NK = 36$, where $N = 6$ and $K = 6$. By the eigenvalue decomposition on clutter covariance

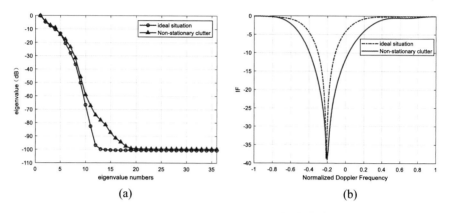

Fig. 2.8 Affection on STAP processing performance by clutter nonstationarity. **a** The eigenvalue spectrum of CCM; **b** Improvement factor in bistatic SAR

matrix, the eigenvalue spectrum of bistatic SAR clutter is demonstrated in Fig. 2.8a. From the result, it can be obviously seen that the eigenvalue of the estimated covariance matrix has increased. This means that by the influence of the clutter nonstationarity, the estimated covariance matrix $\hat{R}_c$ and the real matrix R_c have the different statistical characteristic, which will directly affect the performance of the adaptive filter. Figure 2.8b shows the improvement factor of the adaptive filter in bistatic SAR. Due to the influence of the clutter nonstationarity, the notch of clutter suppression will be significantly widened and an offset will exsit at its center. Therefore, this deterioration of traditional STAP processing performance will impact on the subsequent moving target detection.

Moreover, traditional STAP method will be also influenced by the range migration and Doppler characteristics of the bistatic SAR echo. Because of the characteristics of bistatic SAR, such as the separated transmitter and receiver, the observation state, large bandwidth signals and so on, echo energy will be smeared into multiple range cells and Doppler cells. In addition, the information of range, Doppler frequency and spatial frequency of clutter echo changes with azimuth and time. Therefore, in bistatic SAR, its system and echo characteristics will cause the failure of the "narrowband signal" assumption in traditional STAP method (i.e., without changes of range, angle and Doppler information in one CPI), influencing the clutter suppression effect further.

2.4 Summary

Clutter suppression is an important step for the detection of bistatic SAR moving target, and it affects the detection performance directly. Therefore, this chapter first briefly introduces the basic theory of clutter suppression, including the basic theory of DPCA and STAP. Then, the main issues of bistatic SAR clutter suppression are discussed, including problems of DPCA and traditional STAP in bistatic SAR. For DPCA processing, since the DPCA condition is hard to be satisfied, the phase centers of bistatic SAR receiving channels cannot coincide with each other, which directly limits the performance and application of DPCA. For traditional STAP processing, the non-stationarity of bistatic SAR clutter and its impact are briefly introduced. Through theoretical analysis and simulation, we found that clutter in different range cells had serious dispersion, and strong non-stationarity in bistatic SAR. Therefore, it will seriously affect the clutter suppression performance of traditional STAP. Moreover, the traditional STAP method will be also influenced by the range migration and Doppler characteristics of the bistatic SAR echo. Therefore, to solve the above problems, we proposed several clutter suppression methods in the next chapters to eliminate the influence of bistatic SAR clutter characteristics.

Chapter 3
Bistatic SAR Signal Model

Abstract This chapter first introduces the bistatic SAR time-domain signal model, and then establishes the space–time clutter model for bistatic SAR. This chapter also analyzes the clutter characteristic of bistatic SAR clutter, including the non-stationary characteristic and the space–time distribution with different configuration parameters. The summary of this chapter is shown at the end of this chapter.

Keywords Bistatc SAR clutter · Time-domain signal model · Space–time clutter model · Distribution characteristic · Non-stationary characteristic

3.1 BiSAR Signal Model in Time-Domain

3.1.1 Time-Domain Signal Model

The geometric configuration of bistatic SAR is shown in Fig. 3.1. The receiver is equipped with N channels and the transmitter has only one channel. The original coordinate of the transmitter is (x_T, y_T, z_T). The original coordinate of reference receiving channel is (x_R, y_R, z_R) and the original coordinates of other receiving channels are $(x_R, y_R + (n - 1)d, z_R), n = 2, \ldots, N$, where d is channel spacing. L is the projection of the baseline on the ground between the transmitter and reference receiving channel. V_R and V_T are the velocities of the receiver and the transmitter, respectively. δ_R and δ_T are the flying directions of the receiver and the transmitter, respectively. $\mathbf{V}_R = (0, V_R)$ and $\mathbf{V}_T = (V_{Tx}, V_{Ty})$ are the velocity vectors of the receiver and the transmitter, respectively. $P(x, y)$ donates one arbitrary point target in the observation area. θ_R is the azimuth angle between $P(x, y)$ and the receiver, and θ_T is the azimuth angle between $P(x, y)$ and the transmitter. φ_R is the elevation angle between $P(x, y)$ and the receiver, and φ_T is the elevation angle between $P(x, y)$ and the transmitter. ψ_R and ψ_T represent the cone angles between $P(x, y)$ and platforms. If $P(x, y)$ is a clutter scattering point, its velocity will be $(0, 0)$. If $P(x, y)$ is a potential moving target, its velocity will be (v_x, v_y) and the vector form can be expressed as $\mathbf{V}_{MT}$.

Suppose that the transmitted signal is a linear frequency modulated (LFM) signal. After demodulation, the received echo of the nth receiving channel can be expressed

© The Author(s), under exclusive license to Springer Nature Singapore Pte Ltd. 2022
Z. Li et al., *Bistatic SAR Clutter Suppression*,
https://doi.org/10.1007/978-981-19-0159-1_3

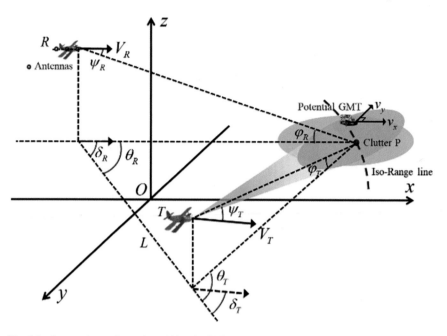

Fig. 3.1 Geometric configuration of bistatic SAR

as

$$S_n(\tau, t; x, y) = \omega_r(\tau - \Delta\tau_n)\exp\{j\pi K_r(\tau - \Delta\tau_n)^2\}$$
$$\times \, \omega_a(t)\exp\{-j2\pi f_c\Delta\tau_n\} \tag{3.1}$$

where $\omega_r(\cdot)$ and $\omega_a(\cdot)$ represent the signal envelopes in range direction and azimuth direction, respectively. τ and t represent the range time and azimuth time. K_r is the range frequency modulated (FM) rate and $f_c = c/\lambda$ is the carrier frequency, where c and λ denote the speed of light and wavelength, respectively. $\Delta\tau_n = R_n(t; x, y)/c$ is the time delay of the received signal, which is reflected back to the nth receiving channel from point $P(x, y)$. $R_n(t; x, y)$ is the bistatic range history of the received echo in the nth receiving channel, which can be expressed as

$$R_n(t; x, y) = R_{Rn}(t; x, y) + R_T(t; x, y) \tag{3.2}$$

where $R_{Rn}(t; x, y)$ is the range history between the nth receiving channel and point $P(x, y)$, and $R_T(t; x, y)$ is the range history between the transmitter and point $P(x, y)$, respectively, which can be expressed as follows

$$R_{Rn}(t; x, y) = \sqrt{(x_R - v_x t - x)^2 + (y_R + (n-1)d + (V_R - v_y)t - y)^2 + z_R^2}$$

(3.3)

$$R_T(t; x, y) = \sqrt{(x_T + (V_{Tx} - v_x)t - x)^2 + (y_T + (V_{Ty} - v_y)t - y)^2 + z_T^2}$$ (3.4)

Therefore, the received signal of $P(x, y)n$ in 2-D time domain is

$$S_n(\tau, t; x, y) = \omega_r\left(\tau - \frac{R_{Rn}(t; x, y) + R_T(t; x, y)}{c}\right)$$

$$\exp\left\{j\pi K_r\left(\tau - \frac{R_{Rn}(t; x, y) + R_T(t; x, y)}{c}\right)^2\right\}$$

$$\times \omega_a(t)\exp\left\{-j2\pi f_c\frac{R_{Rn}(t; x, y) + R_T(t; x, y)}{c}\right\}$$ (3.5)

In the Eq. (3.5), the first exponential term is the range FM signal, which determines the range resolution and the range cell migration (RCM) characteristic in BiSAR. The second exponential term is the azimuth phase of $P(x, y)$, which determines the azimuth FM rate and the Doppler centroid of $P(x, y)$.

After range pulse compression, received signal in range frequency-azimuth time domain is given by

$$S_n(f_\tau, t; x, y) = \omega_r(f_\tau)\omega_a(t)\exp\left\{-j2\pi\frac{f_\tau + f_c}{c}(R_{Rn}(t; x, y) + R_T(t; x, y))\right\}$$

(3.6)

where f_τ denotes range frequency.

In order to analyze and compare the related characteristics (RCM characteristic and the Doppler characteristic) between moving target and clutter, BiSAR range history $R_n(t; x, y)$ can be expressed as

$$R_n(t; x, y) = R_n(0; x, y) + R_n'(0; x, y)t + \frac{1}{2}R_n''(0; x, y)t^2 + \cdots$$ (3.7)

where $R_n(0; x, y)$ is the bistatic range sum of $P(x, y)$ at the initial time. $R_n'(0; x, y)$ and $R_n''(0; x, y)$ are the first-order and the second-order Taylor coefficients of $R_n(t; x, y)$ with respect to t, which can be expressed as

$$R_n(0; x, y) = R_T(0; x, y) + R_{Rn}(0; x, y)$$

$$= R_T(0; x, y) + \sqrt{((n-1)d)^2 + 2(y_R - y)(n-1)d + R_{R0}^2(0; x, y)}$$

(3.8)

$$R'_n(0; x, y) = \frac{((n-1)d + y_R - y)(V_R - v_y) + v_x(x - x_R)}{R_{Rn}(0; x, y)}$$
$$+ \frac{(x - x_T)(v_x - V_{Tx}) + (y - y_T)(v_y - V_{Ty})}{R_T(0; x, y)} \tag{3.9}$$

$$R''_n(0; x, y) = \frac{v_x^2\left(((n-1)d + y_R - y)^2 + z_R^2\right) + \left(z_R^2 + (x - x_R)^2\right)(v_y - V_R)^2}{R_{Rn}^3(0; x, y)}$$
$$+ \frac{2v_x(x - x_R)((n-1)d + y_R - y)(v_y - V_R)}{R_{Rn}^3(0; x, y)}$$
$$+ \frac{\left(z_T^2 + (y - y_T)^2\right)(v_x - V_{Tx})^2 + \left(z_T^2 + (x - x_T)^2\right)(v_y - V_{Ty})^2}{R_T^3(0; x, y)}$$
$$- \frac{2(x - x_T)(y - y_T)(v_x - V_{Tx})(v_y - V_{Ty})}{R_T^3(0; x, y)} \tag{3.10}$$

BiSAR echo characteristic analysis is the foundation of the clutter suppression. In the following, the difference of RCM and Doppler characteristics between moving target and BiSAR clutter scattering point will be analyzed according to the established time domain signal model. Additionally, conclusions are will be made by point target simulations.

3.1.2 RCM Characteristic

Substituting Eqs. (3.8)–(3.10) into (3.6), we have

$$S_n(f_\tau, t; x, y) = \omega_r(f_\tau)\omega_a(t)$$
$$\exp\left\{-j2\pi f_\tau\left(\frac{R_n(0; x, y)}{c} + \frac{R'_n(0; x, y)}{c}t + \frac{1}{2}\frac{R''_n(0; x, y)}{c}t^2 + \cdots\right)\right\}$$
$$\times \exp\left\{-j\frac{2\pi}{\lambda}\left(R_n(0; x, y) + R'_n(0; x, y)t + \frac{1}{2}R''_n(0; x, y)t^2 + \cdots\right)\right\} \tag{3.11}$$

The first exponential term in Eq. (3.11) is the coupling term between range frequency f_τ and azimuth time t. It can be seen that range information of $P(x, y)$ is varied with t and its signal will migrate to multiple range cells during the aperture time.

RCM of $P(x, y)$ in Eq. (3.11) can be expressed as

$$\Delta R_n = R'_n(0; x, y)t + \frac{1}{2}R''_n(0; x, y)t^2 + \cdots \tag{3.12}$$

where the first term is range walk, and the second term is range curvature. In BiSAR systems, range walk and range curvature are the main RCMs. In the following, the RCM characteristic of BiSAR echo will be analyzed. From Eqs. (3.9), (3.10) and (3.12), it can be observed that both range walk and range curvature of $P(x, y)$ are related to its velocity information. Thus, for the targets with different velocities, their RCMs are different. Obviously, moving target's RCM characteristic is different from that of stationary target as well, while their positions are the same at the beam center crossing time.

Figure 3.2 is the simulation results of the range history of one stationary point target with BiSAR configuration. Table 3.1 shows main simulation parameters. Figure 3.2a shows the range history contribution of the transmitter and the receiver, and Fig. 3.2b shows the comparison result of range history between bistatic SAR and monostatic SAR.

As shown in Fig. 3.2a, with the BiSAR configuration in Table 3.1, range walk is the main part in the receiver's range history contribution, resulting in the oblique red line in Fig. 3.2a. Since the squint angle of the transmitter is relatively small, range curvature is relatively large in the transmitter's range history contribution,

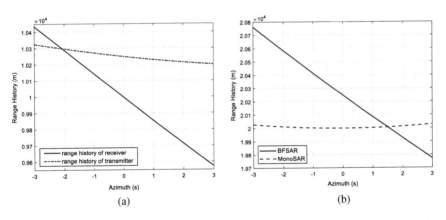

Fig. 3.2 Simulation results of BiSAR range history. **a** Contribution of the receiver and the transmitter; **b** Range history comparison between bistatic SAR and monostatic SAR

Table 3.1 Simulation parameters

Parameters	Values
Transmitter position	$(-8500, -1200, 5600)$ m
Receiver position	$(0, -8000, 6000)$ m
Transmitter velocity	$(0, 180, 0)$ m/s
Receiver velocity	$(0, 180, 0)$ m/s
Receiver detection range	10,000 m
Target center in bistatic SAR	$(0, 0, 0)$ m
Target center in monostatic SAR	$(-8000, -8000, 0)$ m

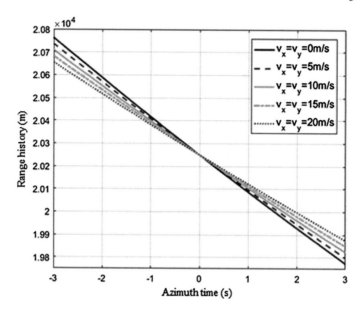

Fig. 3.3 Bistatic range histories of targets with different velocities

resulting in the blue curve in Fig. 3.2a. The range history of BiSAR is the sum of the transmitter's contribution and the receiver's contribution. Due to the great influence of linear terms, range history is generally presented as an oblique line, as shown in Fig. 3.2b. Besides, the linear term in BiSAR range history is much larger than that in monostatic SAR, which will lead to more serious range walk.

Figure 3.3 shows bistatic range histories of targets with different velocities. It is obvious that the RCM characteristic is influenced by target velocity. For the targets with different velocities, their bistatic range history lines have different declines and bendings, as shown in Fig. 3.3. Meanwhile, there is an obvious difference of bistatic range history between stationary target and moving target, which matches the above theoretical analysis.

Figure 3.4 shows the real part of moving target echo and stationary target echo. The comparison results prove that echo inclination along azimuth are different between moving target and stationary target. That is to say, they have significantly different range cell migration. This is in accordance with the range history analysis.

3.1.3 Doppler Characteristic

According to the Eq. (3.11), the signal phase of $P(x, y)$ can be expressed as

$$\phi_n(t) = -\frac{2\pi}{\lambda} \left(R_n(0; x, y) + R'_n(0; x, y)t + \frac{1}{2}R''_n(0; x, y)t^2 + \cdots \right) \qquad (3.13)$$

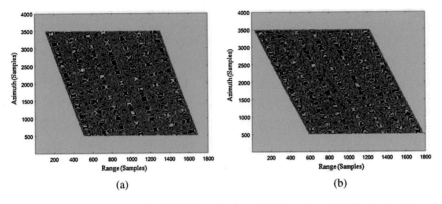

Fig. 3.4 Real part of BiSAR echo in 2-D time domain. **a** Stationary target; **b** Moving target

Thus, the Doppler history of $P(x, y)$ can be calculated as

$$f_{dp}(t) = \frac{1}{2\pi} \frac{d\phi_n(t)}{dt} = -\frac{1}{\lambda}\left(R'_n(0; x, y) + R''_n(0; x, y)t + \cdots\right) = f_{dc} + f_{dr}t + \cdots$$
(3.14)

where f_{dc} and f_{dr} are the Doppler centroid (DC) and DFR of $P(x, y)$ given as follows:

$$f_{dc} = -\frac{R'_n(0; x, y)}{\lambda}$$
(3.15)

$$f_{dr} = -\frac{R''_n(0; x, y)}{\lambda}$$
(3.16)

It can be seen that the Doppler history of $P(x, y)$ is determined by its DC f_{dc} and DFR f_{dr}, as shown in Eq. (3.14). Since f_{dc} and f_{dr} are related to $R'_n(0; x, y)$ and $R''_n(0; x, y)$, Doppler frequency information of $P(x, y)$ is related to its position and velocity. For the moving target and the stationary target, whose positions are the same at the beam center crossing time, their Doppler histories are entirely different. If they have the same Doppler history, their locations will be different. Besides, due to the existence of DFR, the Doppler spectrum of $P(x, y)$ will be broadened in the aperture time. Therefore, the center frequency and broadening degree of Doppler spectrum of moving target and stationary target will be different as well, because of their difference in Doppler parameters.

Figure 3.5 gives the Doppler history comparison of moving target and stationary target. The simulation parameters are given in Table 3.1. The carrier frequency is 10 GHz. As shown in Fig. 3.5a, the movement of moving target will bring extra DC and DFR, which will make the difference on Doppler characteristic between itself and

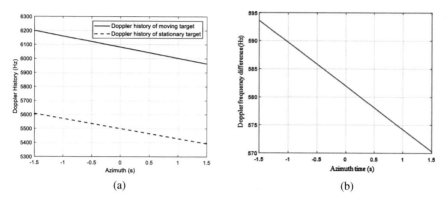

(a)   (b)

Fig. 3.5 Doppler history simulation. **a** Comparison between moving target and stationary target; **b** Doppler history difference between these two target

stationary target. The maximum Doppler difference under this BiSAR configuration will exceed 590 Hz, as depicted in Fig. 3.5b.

Figure 3.6 gives the Doppler spectrums of moving target and stationary target. It can be seen that two Doppler spectrums are broadened. Due to moving target's movement, its DC and Doppler spectrum are obviously different from those of stationary target, which is consistent with the analysis of Doppler history.

Through the above analysis, in BiSAR, there are obvious differences in RCM and Doppler characteristic between stationary target and moving target, and these two characteristics will bring great difficulties in BiSAR clutter suppression. RCM will cause the target energy dispersed in multiple range cells, which will reduce the SCNR, but its influence can be eliminated by the existing RCM correction methods. The broadening of the Doppler spectrum will invalidate the hypothesis of "narrowband" signal in traditional clutter suppression methods, which leads to performance loss in separating clutter and moving target, but its influence can be eliminated by the time division processing proposed in subsequent chapters.

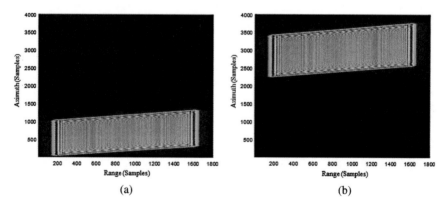

(a)   (b)

Fig. 3.6 Doppler spectrum. **a** Stationary target; **b** Moving target

3.2 BiSAR Clutter Model in Space–Time Domain

Based on the difference of Doppler frequency and spatial position between moving target and stationary target, this section will introduce the coupling relationship between spatial frequency and Doppler frequency of BiSAR clutter, and establish its space–time model, which provides the theoretical support for clutter suppression [46]. Generally, space–time clutter modelling is dependent on the geometrical relationship between ground scattering points and radar platforms. Thus, BiSAR geometrical model will be established first, and then space–time model of BiSAR clutter can be derived.

3.2.1 Geometrical Modelling for BiSAR

In BiSAR, clutter echo of a particular range cell is returned from plenty of clutter patches on the ground. These clutter patches are at the same distance from the receiving and transmitting platforms, and their collection forms an ellipse on the ground. Due to the flexibility of BiSAR platforms in real applications, The flight direction and position of the receiver and transmitter are considered to be arbitrary. Thus, the ellipse constituted by clutter patches is generally nonstandard in BiSAR cases, which greatly increases the difficulty in acquiring the geometrical relationship between ground scattering points and platforms. This section will use the geometrical modelling method based on non-standard ellipse coordinate solution to obtain the geometrical relationship in BiSAR before the precise space–time clutter modelling.

In the Cartesian coordinate system, the nonstandard ellipse can be obtained by rotating and translating its corresponding standard one. Thus, in order to obtain the coordinates of the nonstandard ellipse, we can calculate the coordinates of the standard one first, and then rotate and transform to obtain the coordinates of nonstandard ellipse. That is to say, when geometry center, inclination angle of major axis, major semi-axis and minor semi-axis are defined, the geometrical relationship between scattering points and platforms can be constructed. The derivation process is shown in Fig. 3.7.

Supposing the coordinates of scattering points on the non-standard ellipse are all expressed as (x_0, y_0), the general non-standard ellipse equation is given by

$$ax_0^2 + bx_0y_0 + cy_0^2 + dx_0 + ey_0 + 1 = 0 \qquad (3.17)$$

After operating with vector $(\Delta x, \Delta y)$, the coordinates of scattering points are all expressed as (x_1, y_1) and the geometry center of the new ellipse is located at $(0, 0)$. After rotating angle θ_r, the coordinates of scattering points are all expressed as (x_2, y_2) and the standard ellipse equation is given by

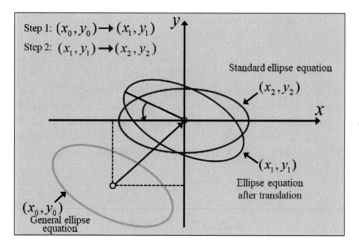

Fig. 3.7 Non-standard ellipse coordinate solution

$$\frac{x_2^2}{L_{ma}^2} + \frac{y_2^2}{L_{mi}^2} = 1, (L_{ma} > L_{mi} > 0) \tag{3.18}$$

where L_{ma} and L_{mi} respectively represent the major semi-axis and the minor semi-axis of standard ellipse, namely the major semi-axis and the minor semi-axis of nonstandard ellipse.

According to the translation of the non-standard ellipse, we have

$$\begin{bmatrix} x_0 \\ y_0 \end{bmatrix} = \begin{bmatrix} x_1 \\ y_1 \end{bmatrix} - \begin{bmatrix} x_p \\ y_p \end{bmatrix} = \begin{bmatrix} x_1 - x_p \\ y_1 - y_p \end{bmatrix} \tag{3.19}$$

Substituting (3.19) into (3.17), the non-standard ellipse equation in matrix representation can be expressed as

$$\begin{bmatrix} a \\ b \\ c \\ d - 2a\Delta x - b\Delta y \\ e - b\Delta x - 2c\Delta y \\ a\Delta x^2 + b\Delta x \Delta y + c\Delta y^2 - d\Delta x - e\Delta y + 1 \end{bmatrix}^T \begin{bmatrix} x_1^2 \\ x_1 y_1 \\ y_1^2 \\ x_1 \\ y_1 \\ 1 \end{bmatrix} = 0 \tag{3.20}$$

After translation, the geometry center of the translational ellipse is $(0, 0)$. That is to say, the coefficients of x_1 and y_1 should be zero. Thus, the translation vector $(\Delta x, \Delta y)$ is satisfied the following equations:

$$\begin{cases} d - 2a\,\Delta x - b\,\Delta y = 0 \\ e - b\,\Delta x - 2c\,\Delta y = 0 \end{cases} \tag{3.21}$$

Therefore, the geometry center of the non-standard ellipse can be obtained by

$$\begin{bmatrix} x_c \\ y_c \end{bmatrix} = \begin{bmatrix} 0 \\ 0 \end{bmatrix} - \begin{bmatrix} \Delta x \\ \Delta y \end{bmatrix} = \begin{bmatrix} \frac{be-2cd}{4ac-b^2} \\ \frac{bd-2ae}{4ac-b^2} \end{bmatrix} \tag{3.22}$$

Similarly, according to the rotation of the translational ellipse, we have

$$\begin{bmatrix} x_1 \\ y_1 \end{bmatrix} = \begin{bmatrix} \cos\theta_r & \sin\theta_r \\ -\sin\theta_r & \cos\theta_r \end{bmatrix} \begin{bmatrix} x_2 \\ y_2 \end{bmatrix} = \begin{bmatrix} \cos\theta_r x_2 + \sin\theta_r y_2 \\ -\sin\theta_r x_2 + \cos\theta_r y_2 \end{bmatrix} \tag{3.23}$$

The rotated ellipse equation in matrix representation can be expressed as

$$\begin{bmatrix} a\cos^2\theta_r + c\sin^2\theta_r - b\cos\theta_r\sin\theta_r \\ 2(a-c)\cos\theta_r\sin\theta_r + b(\cos^2\theta_r - \sin^2\theta_r) \\ a\sin^2\theta_r + c\cos^2\theta_r + b\cos\theta_r\sin\theta_r \\ 0 \\ 0 \\ -ax_c^2 - bx_c y_c - cy_c^2 + 1 \end{bmatrix}^T \begin{bmatrix} x_2^2 \\ x_2 y_2 \\ y_2^2 \\ x_2 \\ y_2 \\ 1 \end{bmatrix} = 0 \tag{3.24}$$

After rotation, the inclination angle of the rotated ellipse is $0°$. That is to say, the coefficients of the coupling term $x_2 y_2$ in Eq. (3.24) should be zero. The rotation angle θ_r is satisfied the following equation:

$$2(a-c)\cos\theta_r\sin\theta_r + b(\cos^2\theta_r - \sin^2\theta_r) = 0 \tag{3.25}$$

The rotation angle θ_r could be obtained by using double angle formula. The inclination angle of the non-standard ellipse can be expressed as

$$\theta_c = -\theta_r = \frac{1}{2}\arctan\left(\frac{b}{a-c}\right) \tag{3.26}$$

After translation and rotation, the corresponding equation of the standard ellipse can be expressed as

$$\begin{bmatrix} a\cos^2\theta_r + c\sin^2\theta_r - b\cos\theta_r\sin\theta_r \\ 0 \\ a\sin^2\theta_r + c\cos^2\theta_r + b\cos\theta_r\sin\theta_r \\ 0 \\ 0 \\ -ax_c^2 - bx_c y_c - cy_c^2 + 1 \end{bmatrix}^T \begin{bmatrix} x_2^2 \\ x_2 y_2 \\ y_2^2 \\ x_2 \\ y_2 \\ 1 \end{bmatrix} = 0 \tag{3.27}$$

Comparing the Eq. (3.27) with (3.18), the major and minor semi-axis of the non-standard ellipse can be expressed as

$$L_{ma} = \sqrt{\frac{ax_c^2 + bx_c y_c + cy_c^2 - 1}{a \cos^2 \theta + c \sin^2 \theta + b \cos \theta \sin \theta}} \tag{3.28}$$

$$L_{mi} = \sqrt{\frac{ax_c^2 + bx_c y_c + cy_c^2 - 1}{a \sin^2 \theta + c \cos^2 \theta - b \cos \theta \sin \theta}} \tag{3.29}$$

Note that all geometrical parameters of the non-standard ellipse have been calculated. With these geometrical parameters, the coordinates of scattering points on the ground can be preciously obtained in any BiSAR configuration.

For the bistatic range R_s, the relationship between instantaneous bistatic range and ground scattering point $P(x, y)$ can be expressed as

$$R_s = \sqrt{(x - x_R)^2 + (y - y_R)^2 + (z - z_R)^2} + \sqrt{(x - x_T)^2 + (y - y_T)^2 + (z - z_T)^2} \tag{3.30}$$

Expand the Eq. (3.30) to get the general ellipse expression:

$$\left[(x_R - x_T)^2 - R_s^2\right]x^2 + 2(x_R - x_T)(y_R - y_T)xy + \left[(y_R - y_T)^2 - R_s^2\right]y^2$$
$$\left[A(x_R - x_T) + 2R_s^2 x_R\right]x + \left[A(y_R - y_T) + 2R_s^2 y_R\right]y + \left[\frac{A^2}{4} - R_s^2 M_R^2\right] = 0 \tag{3.31}$$

where $A = x_T^2 + y_T^2 + z_T^2 - M_R^2 - R_s^2$ and $M_R^2 = x_R^2 + y_R^2 + z_R^2$.

Comparing Eq. (3.31) with (3.17), the coefficients of the general ellipse can be obtained by

$$\begin{cases} a = [4(x_R - x_T)^2 - 4R_s^2]/[A^2 - 4R_s^2(x_R^2 + y_R^2 + z_R^2)] \\ b = 8(x_R - x_T)(y_R - y_T)/[A^2 - 4R_s^2(x_R^2 + y_R^2 + z_R^2)] \\ c = [4(y_R - y_T)^2 - 4R_s^2]/[A^2 - 4R_s^2(x_R^2 + y_R^2 + z_R^2)] \\ d = [4A(x_R - x_T) + 8R_s^2 x_R]/[A^2 - 4R_s^2(x_R^2 + y_R^2 + z_R^2)] \\ e = [4A(y_R - y_T) + 8R_s^2 y_R]/[A^2 - 4R_s^2(x_R^2 + y_R^2 + z_R^2)] \end{cases} \tag{3.32}$$

According to the obtained geometry center (x_c, y_c), inclination angle θ_c, major semi-axis L_{ma} and minor semi-axis L_{mi}, the coordinate of any scattering point $P(x, y)$ with the bistatic range R_s can be calculated. The parametric equation of corresponding standard ellipse is represented as

$$\begin{cases} x_2 = L_{ma} \cos \alpha \\ y_2 = L_{mi} \cos \alpha \end{cases} \tag{3.33}$$

where $\alpha \in [0, 2\pi)$.

After translation and rotation transformation, BiSAR clutter scattering point coordinates can be expressed as

$$\begin{bmatrix} x \\ y \end{bmatrix} = \begin{bmatrix} \cos \theta_c & \sin \theta_c \\ -\sin \theta_c & \cos \theta_c \end{bmatrix} \begin{bmatrix} x_2 \\ y_2 \end{bmatrix} + \begin{bmatrix} x_c \\ y_c \end{bmatrix} \tag{3.34}$$

With the coordinates in Eq. (3.34), the geometrical relationship between clutter scattering point and platform can be determined. The azimuth angle and elevation angle of the transmitter and the receiver are given by

$$\theta_R = \arccos \frac{\mathbf{L}_{R_0 T_0} \cdot \mathbf{L}_{R_0 P}}{\|\mathbf{L}_{R_0 T_0}\| \|\mathbf{L}_{R_0 P}\|} \tag{3.35}$$

$$\theta_T = \arccos \frac{\mathbf{L}_{R_0 T_0} \cdot \mathbf{L}_{T_0 P}}{\|\mathbf{L}_{R_0 T_0}\| \|\mathbf{L}_{T_0 P}\|} \tag{3.36}$$

$$\varphi_R = \arcsin \frac{z_{R,T}}{\|\mathbf{L}_{RP}\|} \tag{3.37}$$

$$\varphi_T = \arcsin \frac{z_{R,T}}{\|\mathbf{L}_{TP}\|} \tag{3.38}$$

where $\|\cdot\|$ is l_2 norm, and $R_0 = (x_R, y_R, 0)$ and $T_0 = (x_T, y_T, 0)$ are projections of the transmitter and the receiver on the ground, respectively. The vectors in Eqs. (3.35)–(3.38) can be expressed as

$$\begin{cases} \mathbf{L}_{R_0 T_0} = (x_T - x_R, y_T - y_R, 0) \\ \mathbf{L}_{RP} = (x - x_R, y - y_R, -z_R) \\ \mathbf{L}_{TP} = (x - x_T, y - y_T, -z_T) \\ \mathbf{L}_{R_0 P} = (x - x_R, y - y_R, 0) \\ \mathbf{L}_{T_0 P} = (x - x_T, y - y_T, 0) \end{cases} \tag{3.39}$$

So far, the geometrical model of BiSAR has been completely established, which lay the foundations for space–time clutter modelling with BiSAR configuration.

3.2.2 Space–Time Clutter Modelling for BiSAR

Space–time clutter modelling means finding out the relationship between the Doppler frequency and spatial frequency of clutter in a range cell, i.e., to obtain the space–time information of clutter spectrum. The normalized Doppler frequency and spatial frequency of the clutter scattering point of BiSAR systems can be expressed as [46]

$$f_{dp} = f_{d0} \cdot T_r = \frac{V_T}{\lambda f_r} \cos \psi_T + \frac{V_R}{\lambda f_r} \cos \psi_R \qquad (3.40)$$

$$f_{sp} = \frac{\Delta R}{\lambda} = \frac{d}{\lambda} \cos \theta_R \cos \varphi_R \qquad (3.41)$$

where

$$\cos \psi_R = \cos \varphi_R \cos(\theta_R - \delta_R) \qquad (3.42)$$

$$\cos \psi_T = \cos \varphi_T \cos(\theta_T - \delta_T) \qquad (3.43)$$

From Eqs. (3.40) to (3.43), angles $\theta_R, \theta_T, \varphi_R$ and φ_T can be obtained by the established geometrical model in Sect. 3.2.1. In addition, platforms' velocities (V_R and V_T), flight directions (δ_R and δ_T), wavelength and PRF are usually known configuration parameters. Thus, the relationship between the Doppler frequency and spatial frequency of BiSAR clutter can be expressed as

$$f_{dp} = \frac{V_T}{\lambda f_r} \cos(\theta_T - \delta_T) \cos \varphi_T + \frac{V_R}{d \cdot f_r} \frac{\cos(\theta_R - \delta_R)}{\cos \theta_R} f_{sp} \qquad (3.44)$$

The relationship in Eq. (3.44) represents the clutter energy locus in space–time domain of the particular range cell. The locus is referred to as the clutter ridge. It can be observed that f_{dp} is no longer proportional to f_{sp} and their relationship is influenced by the transmitter, which is different from monostatic SAR. Thus, the clutter ridge in BiSAR will become a curve rather than a straight line in space–time domain. Moreover, the clutter ridge is closely related to $\cos \varphi_T$, which varies with bistatic range R_s. As a result, ground clutter is range-dependent in BiSAR, i.e., its space–time characteristic varies with bistatic ranges, which will directly impact the estimated performance of CCM in traditional STAP.

Similarly, the space–time model of moving target can be established. The normalized Doppler frequency and spatial frequency of moving target can be expressed as

$$f_{dMT} = \frac{\mathbf{u}_{RM} \cdot (\mathbf{V}_R - \mathbf{v}_{MT}) + \mathbf{u}_{TM} \cdot (\mathbf{V}_T - \mathbf{v}_{MT})}{\lambda f_r} \qquad (3.45)$$

$$f_{sMT} = \frac{d}{\lambda}(\mathbf{u}_{RM} \cdot \mathbf{d}_n) \qquad (3.46)$$

where terms $\mathbf{V}_R$, $\mathbf{V}_T$ and $\mathbf{v}_{MT}$ are the velocity vectors of receiver, transmitter and moving target, respectively. Term $\mathbf{d}_n$ is a unit vector along the array orientation. Terms $\mathbf{u}_{RM}$ and $\mathbf{u}_{TM}$ are unit vectors in the directions from moving target to the transmitter and receiver, respectively, i.e.,

$$\mathbf{u}_{RM} = \frac{\mathbf{L}_{MT} - \mathbf{L}_R}{\|\mathbf{L}_{MT} - \mathbf{L}_R\|} \qquad (3.47)$$

$$\mathbf{u}_{TM} = \frac{\mathbf{L}_{MT} - \mathbf{L}_T}{\|\mathbf{L}_{MT} - \mathbf{L}_T\|} \qquad (3.48)$$

where terms $\mathbf{L}_R$, $\mathbf{L}_T$ and $\mathbf{L}_{MT}$ are the position vectors of receiver, transmitter and moving target, respectively.

3.3 BiSAR Clutter Characteristic Analysis

This section mainly researches and analyzes the nonstationary characteristic of BiSAR clutter and the influence of BiSAR configuration.

As shown in Eq. (3.44), BiSAR clutter ridge represents as a curve in the space–time domain, which is related to bistatic range. Thus, BiSAR clutter distribution in different range cell will be different in space–time domain. BiSAR clutter in different range cell cannot satisfy the I.I.D. requirement and it's nonstationary. Thus, the accuracy of CCM estimation will be influenced and the clutter suppression performance of the traditional method will be degraded.

According to the research and summary of Klemm [47], clutter characteristic analysis is carried out with four typical BiSAR configurations, following flying mode (FFM), parallel flying model (PFM), cross flying mode (CFM) and vertical flying mode (VFM). Simulated parameters are shown in Table 3.2 and simulation results are depicted from Figs. 3.8, 3.9, 3.10 and 3.11. From the simulated results, it can be

Table 3.2 Typical bistatic SAR configuration parameters

	Parameters			
Configurations	Transmitter position (km)	Receiver position (km)	Transmitter velocity (m/s)	Receiver velocity (m/s)
FFM	$(8, -6, 6)$	$(0, -6, 6)$	$(140, 0, 0)$	$(140, 0, 0)$
PFM			$(0, 140, 0)$	$(0, 140, 0)$
CFM			$(100, 100, 0)$	$(0, 140, 0)$
VFM			$(140, 0, 0)$	$(0, 140, 0)$

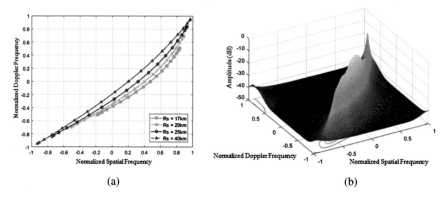

(a)   (b)

Fig. 3.8 Bistatic SAR clutter characteristic simulation in FFM case. **a** Clutter ridges; **b** Clutter spectrum distribution

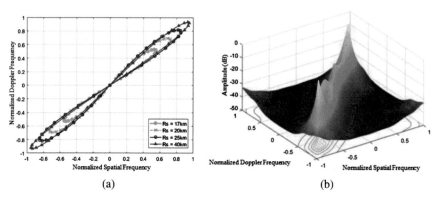

(a)   (b)

Fig. 3.9 Bistatic SAR clutter characteristic simulation in PFM case. **a** Clutter ridges; **b** Clutter spectrum distribution

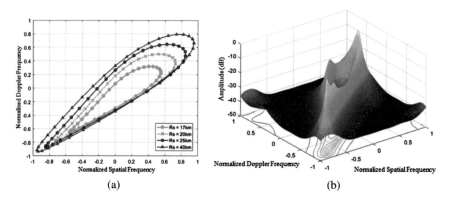

(a)   (b)

Fig. 3.10 Bistatic SAR clutter characteristics simulation in CFM case. **a** Clutter ridges; **b** Clutter spectrum distribution

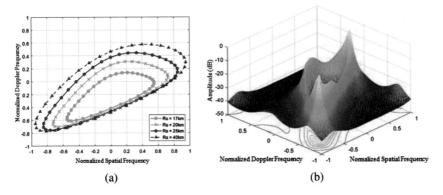

Fig. 3.11 Bistatic SAR clutter characteristics simulation in VFM case. **a** Clutter ridges; **b** Clutter spectrum distribution

seen that BiSAR clutter characteristic is closely related to the locations of transmitter and receiver, speed, course and other configuration parameters.

From Figs. 3.8, 3.9, 3.10 and 3.11, we can observe that BiSAR clutter has strong non-stationarity under different configurations, clutter ridge varies with bistatic range obviously and the clutter spectrum is broadened in different degrees. In PFM case, the clutter ridge looks like a number "8". In the observation area beyond zero Doppler, clutter ridge is still varied with bistatic range. In CFM and VFM cases, clutter ridge has become a closed curve. The shape and varied range of BiSAR clutter distribution under different bistatic ranges have great difference, and the non-stationarity is strong. Through the above analysis, the space–time characteristic of BiSAR clutter is closely related to its configuration parameters, the clutter characteristic is quite different under different BiSAR configurations, and clutter spectrum has different degrees of broadening and diffusion in space–time domain, which brings great difficulties for clutter suppression and moving target detection.

3.4 Summary

This chapter first establishes BiSAR signal model in 2-D time domain, and based on this signal model, analyzes and simulates the RCM and Doppler characteristic of BiSAR signal. The simulation result shows that moving target and clutter have different RCM in BiSAR system. Moving target and clutter at the same original location have different instantaneous Doppler, whereas the spatial locations are different for moving target and clutter when they have the same instantaneous Doppler, which provides a way to separate moving target and background clutter. Then, on the basis of the flexible configuration of BiSAR, the geometrical model is established by using non-standard ellipse coordinate solution method and the space–time clutter model under arbitrary BiSAR configurations is derived as well. At last, with the space–time clutter model, the nonstationary characteristic of BiSAR clutter is analyzed and simulated. The simulation results show that BiSAR clutter is greatly non-stationary. The space–time distribution of clutter in different range cells is significantly different, and its non-stationarity are related to the platform locations, velocity, flight direction and other configuration parameters. All these studies will make a theoretical foundation for the further BiSAR clutter suppression method design and research.

Chapter 4
DPCA-Based Clutter Suppression Method

Abstract This chapter introduces the DPCA-based method for BiSAR clutter suppression. Two kinds of DPCA-based methods are shown in this chapter. The multi-pulse DPCA clutter suppression method is constructed in echo domain, and it mainly includes range cell migration correction, Doppler parameter equalization and multi-pulse canceller construction. Then, image-domain DPCA clutter suppression method is given, including range Doppler-based DPCA (RD-DPCA) and back projection-based DPCA (BP-DPCA). These two kinds of DPCA method are not affected by the nonstationary aspect of BiSAR clutter and they do not need to satisfy the strict DPCA condition.

Keywords Bistatic SAR · Clutter suppression · DPCA · Doppler parameter equalization · Multi-pulse cancellation · Range Doppler · Back Projection

4.1 Multi-pulse DPCA Clutter Suppression Method

This section introduces a clutter suppression method based on multi-pulse DCPA cancellation. Firstly, the range cell migration (RCM) is corrected by first-order Keystone transform [72, 73] and high-order range migration correction function, so that the space–time two-dimensional processing of the echo is realized. Then, spatial spectrum extension elimination and space–time coupling suppression are realized through spatial deramp processing. At the same time, the nonlinear chirp scaling transform (NLCS) is used to equalize Doppler parameters of clutter [74, 75], and the azimuth deramp function is designed to achieve Doppler spectrum extension elimination, so that moving target and clutter scattering points are separated from each other in space–time domain. Finally, a multi-pulse DCPA clutter eliminator is designed to suppress clutter, and the method is verified by simulations.

© The Author(s), under exclusive license to Springer Nature Singapore Pte Ltd. 2022   47
Z. Li et al., *Bistatic SAR Clutter Suppression*,
https://doi.org/10.1007/978-981-19-0159-1_4

4.1.1 Range Cell Migration Correction

Due to the Doppler ambiguity in the echo, it is impossible to directly perform the first-order Keystone transform to correct the range walk of the echo [72, 73]. Therefore, it is necessary to design a pre-filter to remove the Doppler ambiguity of clutter. The pre-filter can be expressed as:

$$h_{pre} = \exp\left\{ j2\pi f_{d0} \frac{(f_\tau + f_c)}{f_c} \eta \right\} \tag{4.1}$$

where f_{d0} is the Doppler centroid of the reference point, which can be expressed as:

$$f_{d0} = \frac{v_r \cos(\theta_R) + v_t \cos(\theta_T)}{\lambda} \tag{4.2}$$

where $\cos(\theta_R)$ and $\cos(\theta_T)$ can be obtained by the following formulas

$$\cos(\theta_R) = \frac{|y_r|}{R_r(0, 1; 0, 0)} \tag{4.3}$$

$$\cos(\theta_T) = \frac{|v_t|^2 + R_t^2(0; 0, 0) - |(x_t + v_{tx}, y_t + v_{ty}, h_t)|^2}{2|v_t|R_t(0; 0, 0)} \tag{4.4}$$

where $|\cdot|$ represents the modulus operation. When moving target velocity is too large, the Doppler ambiguity number of moving target and clutter will be different, so that Eq. (4.1) cannot remove the Doppler ambiguity of moving target and clutter at the same time.

In this case, Doppler ambiguity number of moving target can be estimated based on Radon transform correlation method [76], and then the corresponding pre-filter can be reconstructed to remove Doppler ambiguity.

Signal after removing the Doppler ambiguity can be expressed as

$$\varphi_2(f_\tau, \eta, i; x_P, y_P) = -\frac{2\pi(f_c + f_\tau)}{c} \left\{ \begin{array}{l} R(0, 1; x_P, y_P) + (R'_\eta - \lambda f_{d0})\eta + R'_i i \\ + \frac{1}{2}R''_\eta \eta^2 + \frac{1}{2}R''_i i^2 + R''_{\eta i} \eta i + \frac{1}{6}R'''_\eta \eta^3 \end{array} \right\} \tag{4.5}$$

From Eq. (4.5), it can be seen that RCM existing in the echo and the range walk is the main part of RCM. Range walk can be corrected by the first-order Keystone transform. The Keystone transform can be expressed as

$$\eta = \frac{f_c \eta_1}{(f_c + f_\tau)} \tag{4.6}$$

where η_1 represents a new slow time after the first-order Keystone transform.

Substitute Eq. (4.6) into Eq. (4.5) to obtain a new echo phase, expressed as:

$$\varphi_3\left(f_\tau, \eta_1, i; x_P, y_P\right)$$

$$= -\frac{2\pi}{c} \left\{ \begin{array}{l} (f_\tau + f_c)\left[R(0, 1; x_P, y_P) + R_i' i + \frac{1}{2}R_i'' i^2\right] \\ \\ +f_c\left[R_\eta' - \lambda f_{d0} + R_{\eta i}'' i\right]\eta_1 + \frac{f_c^2}{2(f_c + f_\tau)}R_\eta'' \eta_1^2 + \frac{f_c^3}{6(f_c + f_\tau)^2}R_\eta'' \eta_1^3 \end{array} \right\}$$

$$(4.7)$$

As seen from the above equation, the first-order coupling of f_τ and η_1 is removed, so range walk has been corrected. However, the high-order RCM still exists. $\varphi_3(f_\tau, \eta_1, i; x_P, y_P)$ is obtained by Taylor expansion of f_τ:

$$\varphi_3(f_\tau, \eta, i; x_P, y_P) = -2\pi\frac{f_c}{c}\{R_1(\eta_1, i; x_P, y_P)\}$$

$$-\frac{2\pi}{c}\left\{R(0, 1; x_P, y_P) + R_i' i + \frac{1}{2}R_i'' i^2 - \frac{1}{2}R_\eta'' \eta_1^2 - \frac{1}{3}R_\eta'' \eta_1^3\right\}f_\tau$$

$$-\left\{\frac{\pi}{K_r} + \frac{\pi R_\eta'' \eta_1^2}{c f_c} + \frac{\pi R_\eta'' \eta_1^3}{c f_c}\right\}f_\tau^2 \qquad (4.8)$$

where

$$R_1(\eta_1, i; x_P, y_P)$$

$$= R(0, 1; x_P, y_P) + R_i' i + \frac{1}{2}R_i'' i^2 + (R_\eta' - \lambda f_{d0} + R_{\eta i}'' i)\eta_1 + \frac{1}{2}R_\eta'' \eta_1^2 + \frac{1}{6}R_\eta'' \eta_1^3$$

$$(4.9)$$

According to Eq. (4.8), high-order RCM can be obtained, which can be expressed as

$$RCM_{res}(\eta_1) = \left(-\frac{1}{2}R_\eta'' \eta_1^2 - \frac{1}{3}R_\eta''' \eta_1^3\right)/c \qquad (4.10)$$

In order to correct the high-order RCM, the spatial variation of the high-order RCM needs to be considered first, and the high-order RCM error can be expressed as

$$\Delta R_{num}(\eta_1; x_P, y_P) = RCM_{rec}(\eta_1; x_P, y_P) - RCM_{rec}(\eta_1; 0, 0) \qquad (4.11)$$

where $RCM_{rec}(\eta_1; x_P, y_P)$ and $RCM_{rec}(\eta_1; 0, 0)$ represent the high-order RCM of target and the reference point in the scene, respectively. If $\Delta R_{num}(\eta_1; x_P, y_P)$ is less than a range cell, the higher-order RCM spatial variation can be ignored.

Figure 4.1 is the spatial variation of the high-order RCM. It can be seen from the figure that the maximum difference of high-order RCM in the 2000 m × 2000 m scene is 0.03 range cell, which is far less than one range cell. Therefore, the spatial variation of the high-order RCM can be ignored in the Bistatic SAR. Thus, the corresponding high-order RCM correction function can be expressed as

$$H_{hrcmc}(f_\tau, \eta_1) = \exp\left\{-j\frac{2\pi}{c}\left[\frac{1}{2}R_\eta''(0,0)\eta_1^2 + \frac{1}{3}R_\eta'''(0,0)\eta_1^3\right]f_\tau\right\} \qquad (4.12)$$

where $(0, 0)$ indicates the coordinate of the scene reference point.

The third term in Eq. (4.8) represents the range modulation frequency of echo, and the corresponding range compression function can be expressed as

$$H_{rc}(f_\tau, \eta_1) = \exp\left\{j\left[\frac{\pi}{K_r} + \frac{\pi R_\eta''(0,0)\eta_1^2}{cf_c} + \frac{\pi R_\eta'''(0,0)\eta_1^3}{cf_c}\right]f_\tau^2\right\} \qquad (4.13)$$

Echo signal after RCMC and range compression is expressed as

$$s_1(\tau, \eta_1, i; x_P, y_P)$$
$$= \sigma_0 \exp\{j\varphi_{az}(\eta_1; x_P, y_P)\}\exp\{j\varphi_{aa}(i; x_P, y_P)\}\exp\left(-j\frac{2\pi}{\lambda}R_i''\eta_1 i\right) \qquad (4.14)$$

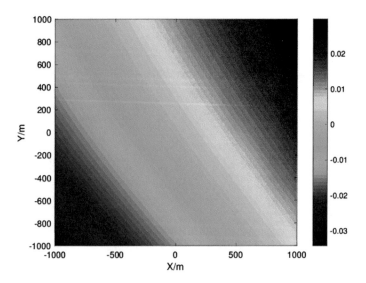

Fig. 4.1 High-order range cell migration spatial variation

where

$$\sigma_0 = \omega_a \left(\frac{\eta - \eta_r}{T_a} \right) \sin c \left\{ B_r \left[\tau - \frac{R(0, 1; x_P, y_P) + R_i' i + \frac{1}{2} R_i'' i^2}{c} \right] \right\} \quad (4.15)$$

$$\varphi_{az}(\eta_1; x_P, y_P) = -\frac{2\pi}{\lambda} \left\{ R(0, 1; x_P, y_P) + \left(R_\eta' - \lambda f_{d0} \right) \eta + \frac{1}{2} R_\eta'' \eta_1^2 + \frac{1}{6} R_\eta''' \eta_1^3 \right\}$$
$$(4.16)$$

$$\varphi_{aa}(i; x_P, y_P) = -j \frac{2\pi}{\lambda} \left(R_i' i + \frac{1}{2} R_i'' i^2 \right) \quad (4.17)$$

In Eq. (4.14), $\varphi_{az}(\eta_1; x_P, y_P)$ and $\varphi_{aa}(i; x_P, y_P)$ represent Doppler phase and spatial phase, respectively. $\exp\left(-j2\pi R_{\eta i}'' \eta_1 i / \lambda\right)$ represents a space–time coupling term. The space–time-range distribution of echo after RCMC is shown in Fig. 4.2. It can be seen that space–time distribution and range domain are separated from each other after RCMC, so that echo processing is reduced from the original space–time-range three-dimensional processing to space–time two-dimensional processing.

4.1.2 Space–Time Spectrum Extension Elimination

From Fig. 4.2b, due to the spatial spectrum extension, Doppler spectrum extension, space–time coupling, moving target and clutter scattering points are overlapped

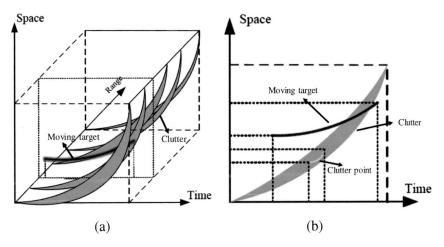

(a) $\qquad\qquad\qquad\qquad\qquad\qquad$ (b)

Fig. 4.2 The distribution of echo after RCMC. **a** The space–time-range distribution of echo; **b** The space–time distribution of the range cell where moving target is located at

with each other in space–time domain, resulting in the failure of effective suppression of clutter. Therefore, in order to separate clutter and moving target in space–time domain, corresponding functions should be designed to eliminate space–time spectrum extension.

First of all, space-variant characteristic of the spatial frequency and space–time coupling term is considered. The phase error of spatial phase error and space–time coupling term can be expressed as

$$\Delta\varphi_d(x_P, y_P; d, i) = -\frac{\pi}{\lambda}\big[R_i''(0, 1; x_P y_P) - R_i''(0, 1; 0, 0)\big]i^2 \tag{4.18}$$

$$\Delta\varphi_c(x_P y_P; d, i, \eta_1) = -\frac{2\pi}{\lambda}\big[R_{\eta i}''(0, 1; x_P, y_P) - R_{\eta i}''(0, 1; 0, 0)\big]\eta_1 i \tag{4.19}$$

where $\Delta\varphi_d(x_P, y_P; d, i)$ is the spatial phase difference caused by spatial variation R_i'', and the space–time coupling phase difference $\Delta\varphi_c(x_P y_P; d, i, \eta_1)$ caused by spatial variation $R_{\eta i}''$.

Figure 4.3 shows the spatial variation of the spatial frequency and space–time coupling term. It can be seen that both the spatial variation of space variation and space–time coupling term are much smaller than 0.25π. Therefore, the spatial variation of the spatial frequency and space–time coupling term can be ignored. Thus, spatial spectrum extension and space–time coupling can be eliminated by a non-space-variable azimuth deramp function. This function can be expressed as

$$H_c(\eta_1, i) = \exp\left\{j\frac{2\pi}{\lambda}\left[R_{\eta i}''(0, 1; 0, 0)\eta_1 i + \frac{1}{2}R_i''(0, 1; 0, 0)i^2\right]\right\} \tag{4.20}$$

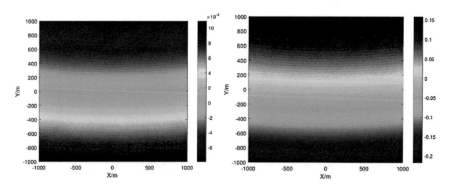

Fig. 4.3 Spatial variation of the spatial frequency and space–time coupling term. **a** Spatial frequency; **b** Space–time coupling term

Fig. 4.4 Space–time distribution after spatial spectrum extension elimination

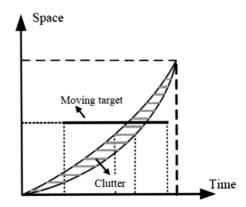

After the above operation, new echo signal can be expressed as:

$$s_2(\tau, \eta_1, i; x_P, y_P) = \sigma_0 \exp\{j\varphi_{az}(\eta_1; x_P, y_P)\} \exp\left\{-j\frac{2\pi}{\lambda}R'_i i\right\} \quad (4.21)$$

Figure 4.4 shows the space–time distribution after spatial spectrum extension elimination and space–time coupling term removal. It can be seen that, due to the existence of Doppler spectrum extension, moving target and clutter scattering points are still aliased with each other in space–time domain, and effective clutter suppression still cannot be achieved.

In order to eliminate the Doppler spectrum extension of echo, the spatial variation characteristic of Doppler parameter is analyzed. According to Eq. (4.16), Doppler phase differences caused by spatial variation of R''_η and R''_η can be expressed as

$$\Delta\varphi_{f2}(x_P, y_P, \eta_1) = -\frac{\pi}{\lambda}\left[R''_\eta(0, 1; x_P, y_P) - R''_\eta(0, 1; 0, 0)\right]\eta_1^2 \quad (4.22)$$

$$\Delta\varphi_{f3}(x_P, y_P, \eta_1) = -\frac{\pi}{3\lambda}\left[R'''_\eta(0, 1; x_P, y_P) - R'''_\eta(0, 1; 0, 0)\right]\eta_1^3 \quad (4.23)$$

where $\Delta\varphi_{f2}(x_P, y_P, \eta_1)$ represents the second-order Doppler phase difference caused by R''_η, and $\Delta\varphi_{f3}(x_P, y_P, \eta_1)$ represents the third-order Doppler phase difference caused by R''_η.

Figure 4.5 shows the analysis result of Doppler spectrum variance. It can be seen that $\Delta\varphi_{f2}(x_P, y_P; \eta_1)$ is much larger than 0.25π and $\Delta\varphi_{f3}(x_P, y_P; \eta_1)$ is much smaller than 0.25π. Thus, the spatial variation of R''_η cannot be ignored, while the spatial variation of R'''_η can be ignored [77].

Since the term R''_η is not negligible, Doppler frequency extension of different clutter points in the same range cell cannot be eliminated by the same azimuth deramp function. To solve this problem, NLCS algorithm can be used to equalize Doppler parameters [75], and then azimuth deramp processing can be used to effectively eliminate the Doppler spectrum extension.

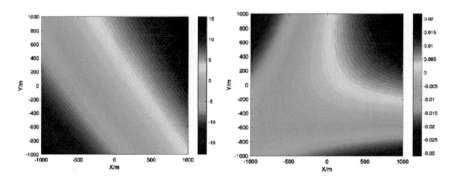

Fig. 4.5 The analysis result of Doppler spectrum variance. **a** Variation of $\Delta\varphi_{f2}(x_P, y_P; \eta_1)$; **b** Variation of $\Delta\varphi_{f3}(x_P, y_P; \eta_1)$

After the first-order Keystone transform, ground clutter points with the same bistatic range are located in the same range cell. According to the relationship between bistatic range and clutter points, the following formula can be obtained

$$x_P(R_0, y_P) = \frac{x_t}{2} + \frac{x_t A}{2B} + \frac{R_0\sqrt{B^2 + A^2 - 2CB}}{2B} \tag{4.24}$$

where

$$A = y_r^2 - y_t^2 + 2y_P(y_t - y_r) + h_r^2 - h_t^2 \tag{4.25}$$

$$B = R_0^2 - x_t^2 \tag{4.26}$$

$$C = (y_P - y_r)^2 + (y_P - y_t)^2 + h_r^2 + h_t^2 \tag{4.27}$$

where R_0 represents the bistatic range.

Substitute Eq. (4.24) into (4.16), at the same time, $\varphi_{az}(\eta_1; x_P, y_P)$ is obtained by Taylor expansion at $\eta_1 = \eta_r$, we have

$$\varphi_{az}(\eta_1; R_0, y_P) = \varphi_{az}(\eta_r; R_0, y_P) + 2\pi f_{dc}(\eta_1 - \eta_r)$$
$$+ \pi f_{dr}(\eta_1 - \eta_r)^2 + \frac{\pi}{3} f_{d3}(\eta_1 - \eta_r)^3 \tag{4.28}$$

where f_{dc}, f_{dr} and f_{d3} represent the Doppler centroid, Doppler frequency rate and third-order Doppler parameters, which can be expressed as

$$f_{dc} = \frac{1}{2\pi} \frac{\partial \varphi_{az}(\eta_1; R_0, y_P)}{\partial \eta_1}\bigg|_{\eta_1=\eta_r} \tag{4.29}$$

$$f_{dr} = \frac{1}{2\pi} \frac{\partial^2 \varphi_{az}(\eta_1; R_0, y_P)}{\partial \eta_1^2} \bigg|_{\eta_1 = \eta_r} \qquad (4.30)$$

$$f_{d3} \approx \frac{1}{2\pi} \frac{\partial^3 \varphi_{az}(\eta_1; R_0, 0)}{\partial \eta_1^3} \bigg|_{\eta_1 = \eta_r} \qquad (4.31)$$

where $\partial/\partial \eta_1$, $\partial^2/\partial \eta_1^2$, and $\partial^3/\partial \eta_1^3$ represent the first, second, and third derivative of η_1, respectively.

The linear fitting of f_{dc} can be obtained by:

$$f_{dc} \approx f_{dc0} + a\eta_r \qquad (4.32)$$

where f_{dc0} represents the Doppler centroid at the initial moment, and a represents the primary fitting coefficient of the Doppler centroid. At the same time, the second-order fitting of f_{dr} can be obtained by:

$$f_{dr} \approx f_{dr0} + b\eta_r + c\eta_r^2 \qquad (4.33)$$

where b and c are the fitting coefficients.

Figure 4.6 shows the fitting results of the Doppler centroid and the Doppler frequency rate. It can be seen that the phase error caused by linear fitting and second-order fitting is much smaller than 0.25π, so it is reasonable to fit the Doppler centroid and Doppler frequency. To simplify subsequent calculations, a compensation function can be set to remove the Doppler centroid of each range cell. The compensation function can be expressed as

$$H_c(\eta_1) = \exp(-j2\pi f_{dc0}\eta_1) \qquad (4.34)$$

After the compensation function, $\varphi_{az}(\eta_1; x_P, y_P)$ can be expressed as

$$\varphi_{az}(\eta_1; R_0, y_P) = \varphi_{az}(\eta_r; R_0, y_P) + 2\pi a\eta_r(\eta_1 - \eta_r)$$
$$+ \pi f_{dr}(\eta_1 - \eta_r)^2 + \frac{\pi}{3} f_{d3}(\eta_1 - \eta_r)^3 \qquad (4.35)$$

After FFT processing, $\varphi_{az}(f_{\eta_1}; x_P, y_P)$ is multiplied by a fourth-order filtering function, which is given by

$$H_{4th-filter}(f_{\eta_1}) = \exp\{j\pi(Y_3 f_{\eta_1}^3 + Y_4 f_{\eta_1}^4)\} \qquad (4.36)$$

where Y_3 and Y_4 are the corresponding coefficients. After the fourth-order filtering function, the azimuth signal can be expressed as

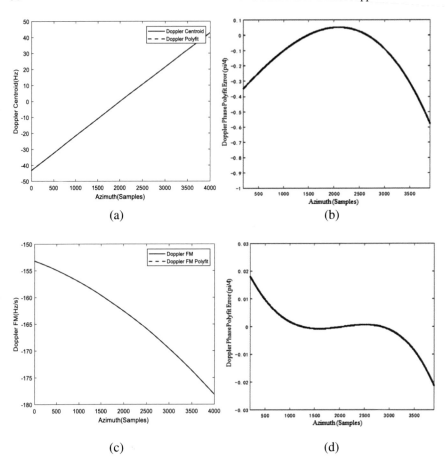

Fig. 4.6 The fitting result of the Doppler centroid and the Doppler frequency rate. **a** Doppler centroid and its linear fitting; **b** Fitting phase error of the Doppler centroid; **c** Doppler frequency rate and its second-order fitting; **d** Fitting phase error of the Doppler frequency rate

$$\varphi_{az}\left(f_{\eta_1}; R_0, y_P\right) = \varphi_{az}(\eta_r; R_0, y_P) - 2\pi f_{\eta_1}\eta_r - \pi \frac{\left(f_{\eta_1} - a\eta_r\right)^2}{f_{dr}}$$

$$+ \frac{\pi f_{d3}}{3} \frac{\left(f_{\eta_1} - a\eta_r\right)^2}{f_{dr}} + \pi\left(Y_3 f_{\eta_1}^3 + Y_4 f_{\eta_1}^4\right) \qquad (4.37)$$

And then transform $\varphi_{az}\left(f_{\eta_1}; R_0, y_P\right)$ into time domain, we have

$$\varphi_{az1}(\eta_1; R_0, y_P)$$

$$= \varphi_{az}(\eta_r; R_0, y_P) + 2\pi a \eta_r(\eta_1 - \eta_r) + \pi f_{dr}(\eta_1 - \eta_r)^2 + \frac{\pi}{3} f_{d3}(\eta_1 - \eta_r)^3$$

$$+ \pi Y_3[f_{dr}(\eta_1 - \eta_r) + a\eta_r]^3 + \pi Y_4[f_{dr}(\eta_1 - \eta_r) + a\eta_r]^4$$

$$+ \varphi_{az}(\eta_r; R_0, y_P) - 2\pi f_{\eta_1}\eta_r - \pi \frac{(f_{\eta_1} - a\eta_r)^2}{f_{dr0} + b\eta_r + c\eta_r^2} \tag{4.38}$$

Next, an azimuth NLCS function can be expressed as

$$H_{NLCS}(\eta_1) = \exp\{j\pi(q_2\eta_1^2 + q_3\eta_1^3 + q_4\eta_1^4)\} \tag{4.39}$$

where q_2, q_3 and q_4 are the corresponding coefficients. The azimuth signal $\varphi_{az1}(\eta_1; R_0, y_P)$ processed by the fourth-order filtering function is multiplied by the azimuth NLCS function. After NLCS processing, the phase in frequency domain can be expressed as

$$\varphi_{az1}(f_{\eta_1}; R_0, y_P)$$

$$= D(f_{\eta_1}) + E\eta_r f_{\eta_1} + F\eta_r^2 f_{\eta_1} + G\eta_r f_{\eta_1}^2 + H\eta_r^2 f_{\eta_1}^2 + I\eta_r f_{\eta_1}^3 + J(\eta_r) \tag{4.40}$$

where D and J represent the coefficients of f_{η_1} and η_r terms respectively. E, G, H and I represent the coefficients of the coupling terms of η_r and f_{η_1}, which can be obtained by the following formulas

$$E = \frac{2\pi(a - f_{dr0})}{q_2 + f_{dr0}} \tag{4.41}$$

$$F = \frac{2\pi b(q_2 - a)}{(q_2 + f_{dr0})^2} + \pi d_{d3}\frac{(a + q_2)^2}{(q_2 + f_{dr0})^3}$$

$$+ 3\pi Y_3\frac{(f_{dr0} - a)^2 a_2^2 f_{dr0}}{(q_2 + f_{dr0})^3} + 3\pi q_3\frac{(a - f_{dr0})^2}{(q_2 + f_{dr0})^3} \tag{4.42}$$

$$G = \frac{\pi b}{(q_2 + f_{dr0})^2} - \pi f_{d3}\frac{(a + q_2)}{(q_2 + f_{dr0})^3}$$

$$+ 3\pi Y_3\frac{(f_{dr0} - a)f_{dr0}^2}{(q_2 + f_{dr0})^3} - 3\pi q_3\frac{(a - f_{dr0})}{(q_2 + f_{dr0})^3} \tag{4.43}$$

$$H = +\pi Y_4\frac{6f_{dr0}^2(f_{dr0} - a)q_2^2}{(q_2 + f_{dr0})^4} + 3\pi q_3\frac{b(q_2 + 3a - 2f_{dr0})}{(q_2 + f_{dr0})^4} + 6\pi q_4\frac{(a - f_{dr0})^2}{(q_2 + f_{dr0})^4}$$

$$\pi \frac{c(q_2 + f_{dr0}) - b^2}{(q_2 + f_{dr0})^3} + 3\pi f_{d3}\frac{b(a + q_2)}{(q_2 + f_{dr0})^4} - 3\pi Y_3 q_2\frac{bf_{dr0}(3f_{dr0}q_2 - 2aq_2 + f_{dr0}a)}{(q_2 + f_{dr0})^3} \tag{4.44}$$

$$I = -\frac{\pi f_{d3} b}{(q_2 + f_{dr0})^4} + 3\pi Y_3 \frac{b f_{dr0}^2 q_2}{(q_2 + f_{dr0})^4} - 4\pi Y_4 \frac{f_{dr0}^3 (f_{dr0} - a) q_2}{(q_2 + f_{dr0})^4}$$

$$-3\pi q_3 \frac{b}{(q_2 + f_{dr0})^4} - 4\pi q_4 \frac{(a - f_{dr0})}{(q_2 + f_{dr0})^4} \tag{4.45}$$

In order to equalize Doppler parameters, the first-order coupling coefficient E of η_r and f_{η_1} is set as $-\pi/\gamma$, where γ determines the azimuth position of clutter points. Other higher-order coupling coefficients are set to be zero. Thus, we have

$$q_2 = -2a\gamma + (2\gamma - 1) f_{dr0} \tag{4.46}$$

$$q_3 = \frac{2b(q_2 + a)(q_2 + f_{dr0}) - f_{d3}(a + q_2)^2 - q_2 N}{3(f_{dr0} - a)^2} \tag{4.47}$$

$$Y_3 = \frac{b(2q_2 + a + f_{dr0}) - f_{d3}(a + q_2)}{3(f_{dr0} - a)^2 q_2 f_{dr0}} \tag{4.48}$$

$$Y_4 = \frac{L/6 - M(a - f_{dr0})/4}{(f_{dr0} - a)^2 q_2 f_{dr0}^2 (q_2 + f_{dr0})} \tag{4.49}$$

$$q_4 = \frac{M/4 - (f_{dr0} - a) f_{dr0}^3 q_2 Y_4}{a - f_{dr0}} \tag{4.50}$$

where

$$N = b(2q_2 + a + f_{dr0}) - f_{d3}(a + q_2) \tag{4.51}$$

$$L = -\left[c(q_2 + f_{dr0})^2 - b^2(q_2 + f_{dr0})\right] - 3 f_{d3} b(a + q_2)$$
$$- 3q_3 b(q_2 - 2 f_{dr0} + 3a) + 3Y_3 q_2 b f_{dr0}(3 f_{dr0} q_2 - 2a q_2 + f_{dr0} a) \tag{4.52}$$

$$M = -3 f_{d3} b + 3 Y_3 q_2 b f_{dr0}^2 - 3q_3 b \tag{4.53}$$

$$\gamma = 0.5 \tag{4.54}$$

After processed with the fourth order filtering function and azimuth NLCS function, the signal phase in the frequency domain can be expressed as

$$\varphi_{az2}(f_{\eta_1}; R_0, y_P)$$
$$\approx -2\pi \eta_r f_{\eta_1} - \frac{\pi f_{\eta_1}^2}{f_{dr0} + q_2} + \pi \frac{(f_{d3}/3 + Y_3 f_{dr0}^3 + q_3) f_{\eta_1}^3}{(q_2 + f_{dr0})^3} + \pi \frac{(Y_4 f_{dr0}^4 + q_4) f_{\eta_1}^4}{(q_2 + f_{dr0})^4} \tag{4.55}$$

It can be seen that the Doppler centroid and the Doppler frequency rate are no longer related to the location of clutter scattering points. That is to say, Doppler parameters are not spatial-variant.

After azimuth NLCS processing, Doppler spectrum extension can be elinimated by the azimuth deramp processing. The third and fourth terms in Eq. (4.55) can be compensated first, and the compensation function can be expressed as

$$H\left(f_{\eta_1}; R_0, y_P\right) = \exp\left\{j\left[\pi\frac{\left(f_{d3}/3 + Y_3 f_{dr0}^3 + q_3\right)f_{\eta_1}^3}{\left(q_2 + f_{dr0}\right)^3} + \pi\frac{\left(Y_4 f_{dr0}^4 + q_4\right)f_{\eta_1}^4}{\left(q_2 + f_{dr0}\right)^4}\right]\right\}$$

(4.56)

After compensation, the echo phase in time domain can be expressed as

$$\varphi_{az3}(\eta_1; R_0, y_P) = \pi(q_2 + f_{dr0})(\eta_1 - \eta_r)^2$$

(4.57)

The azimuth deramp function can be constructed as

$$H_{ac}(\eta_1) = \exp\left[j\pi(q_2 + f_{dr0})\eta_1^2\right]$$

(4.58)

With azimuth deramp processing, echo signal can be expressed as

$$s_3(\tau, \eta_1, i; R_0, y_P)$$
$$= \sigma_1 \exp\left\{-j\frac{2\pi}{\lambda}R_i'i\right\}\exp\{-j2\pi[(q_2 + f_{dr0})\eta_r]\eta_1\}$$

(4.59)

where

$$\sigma_1 = \sigma_0 \exp\left\{j\pi(q_2 + f_{dr0})\eta_r^2\right\}$$

(4.60)

After azimuth deramp processing, space–time spectrum extension is eliminated and the space–time coupling term in echo is removed. Thus, clutter and moving target can be well separated from each other in space–time domain. Figure 4.7 is the echo distribution after space–time spectrum extension elimination. Because of the motion of the target, the residual Doppler spectrum extension still exists after azimuth deramp processing.

4.1.3 Multi-pulse Two-Channel Clutter Cancellation

In thie section, the corresponding clutter canceller is designed to achieve clutter suppression. First, the echo signal can be expressed into matrix form as follows

Fig. 4.7 Echo distribution after space–time spectrum extension elimination

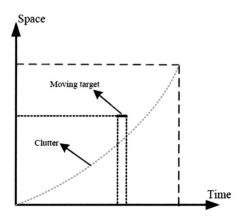

$$s_4(p_1 : p_k, \tau, i) = U V^i \sigma \tag{4.61}$$

where U and V are the Doppler frequency matrix and spatial frequency matrix respectively, which can be expressed as

$$U = \begin{bmatrix} u_{p1}(1) & \cdots & u_{p1}(Q) \\ \vdots & \ddots & \vdots \\ u_{pk}(1) & \cdots & u_{pk}(Q) \end{bmatrix} \tag{4.62}$$

$$V = diag[v(1), \cdots, v(Q)] \tag{4.63}$$

$$\sigma = [\sigma_1(1), \cdots, \sigma_1(1)]^T \tag{4.64}$$

where $diag[\cdot]$ represents diagonal matrix, $[\cdot]^T$ represents transpose operation. Q represents the number of clutter points in a synthetic aperture, which can be expressed as: $Q = L_a/\rho_a$, where $L_a = v_r T_a$ and ρ_a is the azimuth resolution. $u_{pk}(q)$ represents the Doppler phase of clutter point $q \in [1, Q]$ at the azimuth sampling time $p_k p_k \in [p_1, p_k]$, and $v(q)$ represents the spatial phase of clutter point $q \in [1, Q]$. $u_{pk}(q)$ and $v(q)$ can be obtained as follows:

$$u_{pk}(q) = \exp\{-j2\pi(q_2 + f_{dr0})\eta_r(q)\eta_1(p_k)\} \tag{4.65}$$

$$v(q) = \exp\left\{-j\frac{2\pi}{\lambda}R_i'(0, 1; x_q, y_q)\right\} \tag{4.66}$$

where $\eta_r(q)$ represents the beam center crossing time of the q clutter point, and $\eta_1(p_k)$ represents the azimuth sampling moment of the p_k.

Assuming that the weight coefficient of clutter canceller is $Z \in C^{K \times K}$, echo after clutter cancellation can be expressed as

$$s_5(p_1 : p_k, \tau) = Zs_4(p_1 : p_K, \tau, 1) - s_4(p_1 : p_K, \tau, 2)$$
$$= ZUV\sigma - UV^2\sigma = (ZU - UV)V\sigma \qquad (4.67)$$

In order to minimize clutter energy, $\|(ZU - UV)V\sigma\|_F$ needs to be minimized, where $\|\cdot\|_F$ represents the Frobenius norm. According to Cauchy's inequality [78], we can get $\|(ZU - UV)V\sigma\|_F \leq \|(ZU - UV)\|_F \|V\sigma\|_F$. Assuming that $\|V\sigma\|_F$ is a constant and the gradient of the variable Z in $\|(ZU - UV)\|_F$ is minimized, the weight matrix can be obtained by

$$Z = UVU^H \left(UU^H \right)^{-1} \qquad (4.68)$$

It can be seen from Eq. (4.68) that the weight matrix Z is only related to U and V. For different range cells, U and V are only related to the spatial frequency and Doppler frequency of the corresponding range cell [79]. Thus, the weight matrix is range-independent. Therefore, clutter non-stationarity does not affect this method [80], and this method has a better clutter suppression effect than the traditional STAP method in bistatic SAR.

4.1.4   Simulated Analysis

Figure 4.8 shows the geometrical configuration of bistatic SAR system. In this simulation, two moving targets $M1$ and $M2$ are set, with the speeds of $(-3, 4)$ m/s and $(9, -10)$ m/s, respectively. The target RCS are 4 dB and 11 dB, respectively. The RCS of ground grassland and ground building are 5 dB and 20 dB respectively.

It can be seen from the Fig. 4.9 that before the range migration correction, echo in black dashed ellipse is titlting compared with the coordinate axis, and clutter energy is distributed in multiple range cells. After range migration correction, echo in black dashed ellipse is parallel to the azimuth axis, so energy of a single clutter point is

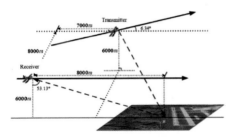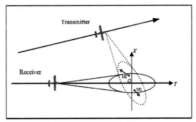

Fig 4.8   The geometrical configuration of bistatic SAR

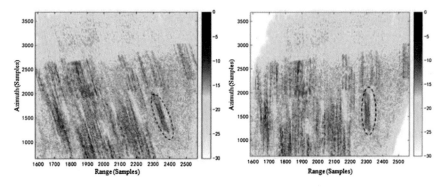

Fig. 4.9 Bistatic SAR echo before and after RCMC. **a** Before RCMC; **b** After RCMC

concentrated in a single range cell. The simulation result in Fig. 4.9 has proved the correctness of range migration correction.

The focusing results of a single point with and without NLCS processing are given in Fig. 4.10. It can be seen that there is no Doppler spectrum extension after NLCS processing, and the point signal is well focused. However, there is still residual Doppler spectrum extension without NLCS processing, and the focusing performance is poor. Therefore, NLCS processing in multi-pulse two-channel cancellation method is necessary and effective for bistatic SAR clutter suppression.

The suppression results of DPCA, STAP and multi-pulse two-channel clutter cancellation are presented in Fig. 4.11. Figure 4.11a is the image of the scene

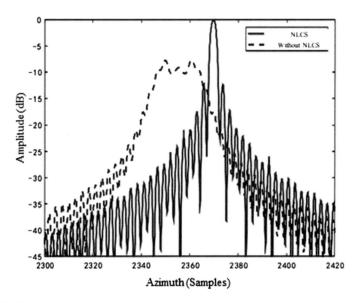

Fig. 4.10 The focusing result of a single point target

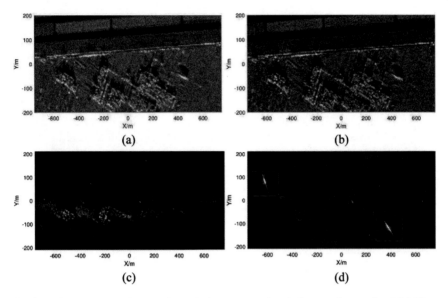

Fig. 4.11 Clutter suppression results. **a** Before suppression. **b** Suppression result of DPCA. **c** Suppression result of STAP. **d** Suppression result of multi-pulse two-channel clutter cancellation

before clutter suppression. It can be seen that due to the low SCNR, moving target is submerged in strong clutter and cannot be effectively detected. From Fig. 4.11b, DPCA method only suppresses a small amount of clutter, and the two moving targets were still submerged in strong clutter. As seen in Fig. 4.11c, clutter suppression effect of STAP is significantly better than that of DPCA method. A large amount of clutter is suppressed, and moving target $M2$ can be detected. However, since the RCS of moving target $M1$ is small, it is still submerged in clutter after STAP method and cannot be effectively detected. From Fig. 4.11d, after multi-pulse two-channel clutter cancellation, clutter is effectively suppressed, and signals of $M1$ and $M2$ are well retained, so that they can be effectively detected. The results in Fig. 4.11 have proved that processing performance of the multi-pulse two-channel clutter cancellation method in bistatic SAR clutter suppression is better than that of traditional DPCA method and STAP method.

4.2 Image-Domain DPCA Clutter Suppression Method

This section will introduce the image-domain DPCA clutter suppression methods, including RD-based DCPA method and BP-based DPCA method.

4.2.1   RD-Based DPCA Method

This subsection will introduce the RD-based DPCA method with two receiving channels. First, the problem of equivalent displaced phase center of two channels in airborne BiSAR is analyzed, deducing the phase displacement formula. In addition, based on the characteristic of phase displacement and the analysis of the phase compensation strategy, the influence on the equivalent phase center caused by the transmitter's movement can be eliminated, through compensating the terms related to clutter's range history and azimuth time. After compensation and FFT, clutter singal can be cancelled in range-Doppler domain, while the information of moving target can be retained after two-channel cancellation, due to its own velocity. Thus, moving target can be further detected after RD-based DPCA processing.

4.2.1.1   Analysis of Two-Channel Equivalent Phase Center

As shown in Fig. 4.12, two receiving channels are installed along-track on the receiver of BiSAR, while one transmitting channel is set on the transmitter. $Y-axis$ represents the direction of platforms, that is, azimuth direction. $X-axis$ represents the direction perpendicular to the flight track, namely, range direction. $Z-axis$ represents the height. The coordinate of the transmitter is (x_T, y_T, h_T) and the speed of the transmitter is V_T. The coordinates of receiving channel 1 and receiving channel 2 are (x_{R1}, y_{R1}, h_{R1}) and (x_{R2}, y_{R2}, h_{R2}), respectively. The distance between two channles is d and the speed of the receiver is V_R. $P_s(x_0, y_0)$ is clutter scattering point. $R_{T,c}$ represents the distance from the transmitter to the clutter point at beam center crossing time, while $R_{R1,c}$ and $R_{R2,c}$ are the distances from receiving channel 1 and channel 2 to the clutter point at beam

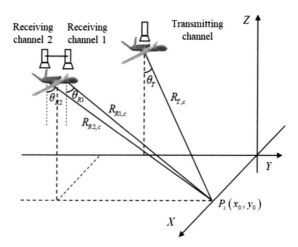

Fig. 4.12 Configuration of airborne BiSAR system with two receiving channels

Fig. 4.13 The range history of bistatic SAR with two receiving channels

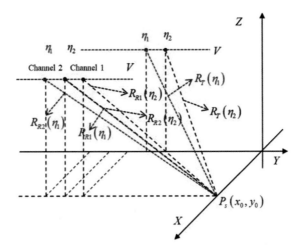

center crossing time. θ_T, θ_{R1} and θ_{R2} are defined as squint angles of the transmitter, receiving channel 1 and channel 2, respectively.

Different from monostatic SAR system, airborne BiSAR operates in mode of "one transmitter and two receivers", making equivalent phase centers of the receiving channels and transmitting channel at different positions in space, due to the motion of the separate transmitter and receiver. As shown in Fig. 4.13, receiving channel 1 at the former azimuth time η_1 and receiving channel 2 at the later azimuth time η_2 are located at the same position. However, since the separate transmitting platform is moving, positions of the transmitter at different azimuth time are different, resulting in the different bistatic range histories in two receiving channels. The problem of non-coincidence of the equivalent phase centers of two channels is analyzed in detail in the following, to build a foundation for the further research of clutter suppression.

In Fig. 4.13, $R_1(\eta_1)$ and $R_2(\eta_1)$ are the range histories between the clutter scattering point at time η_1 and receiving channel 1 and channel 2, respectively. While, $R_1(\eta_2)$ and $R_2(\eta_2)$ are the range histories at time η_2. $R_T(\eta_1)$ is considered as the range history between the clutter scattering point and the transmitter at time η_1, whereas $R_T(\eta_2)$ is the range history contributed by the transmitter at time η_2. Times η_1 and η_2 are satisfied the relationship: $\eta_2 = \eta_1 + m \cdot PRI$, where m is an integer. And the channel spacing d satisfies the relationship: $d = m \cdot PRI$.

At η_1, the bistatic range of receiving channel 1 relative to $P_s(x_0, y_0)$ is

$$R_1(\eta_1, x_0, y_0) = R_T(\eta_1, x_0, y_0) + R_{R1}(\eta_1, x_0, y_0)$$
$$\approx R_{T,c} + R_{R1,c} + (-V_T sin\theta_T - V_R sin\theta_{R1})\eta_1 + \frac{1}{2}\left(\frac{V_T^2 cos^2 \theta_T}{R_{T,c}} + \frac{V_R^2 cos^2 \theta_{R1}}{R_{R1,c}}\right)\eta_1^2$$

$$(4.69)$$

$$\text{where}\begin{cases} R_{T,c} = \sqrt{(x_T - x_0)^2 + (y_T - y_0)^2 + h_T^2} \\ R_{R1,c} = \sqrt{(x_{R1} - x_0)^2 + (y_{R1} - y_0)^2 + h_{R1}^2} \end{cases} \text{and} \begin{cases} \theta_T = \arcsin\left(\frac{y_0 - y_T}{R_{T,c}}\right) \\ \theta_{R1} = \arcsin\left(\frac{y_0 - y_{R1}}{R_{R1,c}}\right) \end{cases}.$$

At η_1, the bistatic range of receiving channel 1 relative to $P_s(x_0, y_0)$ is:

$$R_2(\eta_1, x_0, y_0) = R_T(\eta_1, x_0, y_0) + R_{R2}(\eta_1, x_0, y_0)$$
$$\approx R_{T,c} + R_{R2,c} + (-V_T \sin\theta_T - V_R \sin\theta_{R2})\eta_1 + \frac{1}{2}\left(\frac{V_T^2 \cos^2\theta_T}{R_{T,c}} + \frac{V_R^2 \cos^2\theta_{R2}}{R_{R2,c}}\right)\eta_1^2$$

$$(4.70)$$

$$\text{where}\begin{cases} R_{R2,c} = \sqrt{(x_{R2} - x_0)^2 + (y_{R2} - y_0)^2 + h_{R2}^2} \\ \theta_{R2} = \arcsin\left(\frac{y_0 - y_{R2}}{R_{R2,c}}\right) \end{cases}.$$

After channel 2 moving along the azimuth direction with $\Delta\eta = d/V_R$, bistatic range of channel 2 at η_2 turns to be

$$R_2(\eta_2, x_0, y_0) = R_1(\eta_1 + \Delta\eta, x_0, y_0) = R_T(\eta_1 + \Delta\eta, x_0, y_0) + R_{R2}(\eta_1 + \Delta\eta, x_0, y_0)$$
$$\approx R'_{T,c} + R'_{R2,c} + (-V_T \sin\theta'_T - V_R \sin\theta'_{R2})\eta_1 + \frac{1}{2}\left(\frac{V_T^2}{R'_{T,c}}\cos^2\theta'_T + \frac{V_R^2}{R'_{R2,c}}\cos^2\theta'_{R2}\right)\eta_1^2$$

$$(4.71)$$

$$\text{where}\begin{cases} R'_{T,c} = \sqrt{(x_T - x_0)^2 + (y_T + d - y_0)^2 + h_T^2} \\ R'_{R2,c} = \sqrt{(x_{R2} - x_0)^2 + (y_{R2} + d - y_0)^2 + h_{R2}^2} \end{cases} \text{and}$$

$$\begin{cases} \theta'_T = \arcsin\left(\frac{y_0 - y_T - d}{R_{T,c}}\right) \\ \theta'_{R2} = \arcsin\left(\frac{y_0 - y_{R2} - d}{R_{R2,c}}\right) \end{cases}.$$

Comparing Eqs. (4.69) with (4.71), it can be find that bistatic range of channel 1 at η_1 and channel 2 at η_2 are different, illustrating that the two-channel equivalent phase centers are not coincided with each other in space. At this time, the range difference of two channels can be obtained as follows:

$$\Delta R_{12}(\Delta\eta; x_0, y_0) = R_1(\eta_1, x_0, y_0) - R_2(\eta_2, x_0, y_0)$$
$$= R_{T,c} - R'_{T,c} + R_{R2,c} - R'_{R2,c} + (-V_T \sin\theta_T + V_T \sin\theta'_T)\eta_1 + (-V_R \sin\theta_{R2} + V_R \sin\theta'_{R2})\eta_1$$
$$+ \frac{1}{2}\left(\frac{V_T^2 \cos^2\theta_T}{R_{T,c}} - \frac{V_T^2 \cos^2\theta'_T}{R'_{T,c}}\right)\eta_1^2 + \frac{1}{2}\left(\frac{V_R^2 \cos^2\theta_{R2}}{R_{R2,c}} - \frac{V_R^2 \cos^2\theta'_{R2}}{R'_{R2,c}}\right)\eta_1^2$$

$$(4.72)$$

Since the receiving channel 1 at η_1 and receiving channel 2 at η_2 are located at the same position, relationships of $\theta_{R1} = \theta'_{R2}$ and $R_{R1,c} = R'_{R2,c}$ can be obtained. Then, Eq. (4.72) can be simplified as:

$$\Delta R_{12}(\Delta\eta; x_0, y_0) = \left(R_{T,c} - R'_{T,c}\right) - V_T\left(\sin\theta'_T + \sin\theta_T\right)\eta_1$$
$$+ \frac{1}{2}\left(\frac{V_T^2\cos^2\theta'_T}{R'_{T,c}} - \frac{V_T^2\cos^2\theta_T}{R_{T,c}}\right)\eta_1^2 \tag{4.73}$$

It can be observed that the first term in $\Delta R_{12}(\Delta\eta; x_0, y_0)$ is a constant term, the second term is the first order of azimuth time η_1, and the third term is the second order of azimuth time η_1. These terms are dependent on the parameters of the transmitter. That means that range history difference of two receiving channels is caused by the movement of the transmitting platform, and it also varies with azimuth time. Besides, $R_{T,c}$ and $R'_{T,c}$ are related to the position of the clutter scattering point. Thus, the range history difference also varies with spatial position.

The corresponding phase difference can be expressed as:

$$\Delta\Phi_{12}(\Delta\eta; x_0, y_0) = 2\pi\frac{\Delta R_{12}(\Delta\eta; x_0, y_0)}{\lambda} \tag{4.74}$$

The phase in Eq. (4.74) varies with azimuth time and spatial position of the clutter scattering point as well, which is same with $\Delta R_{12}(\Delta\eta; x_0, y_0)$.

4.2.1.2 Two-Channel Clutter Suppression Based on RD Algorithm

According to the analysis in the previous section, the equivalent phase centers of two channels are not coincided with each other, due to the movement of the separate transmitting platform. The phase difference changes with both the azimuth time and the clutter's position, which is difficult to be compensated. This section will introduce the clutter suppression method based on RD algorithm for bistatic SAR with two receiving channels.

In order to suppress bistatic SAR clutter effectively, the stationary clutter signals received by two channels must be the same, that is, the range history of them must be the same, which satisfies $R_1(\eta_1, x_0, y_0) = R_2(\eta_2, x_0, y_0)$. Thus, the difference of range history of two channels $\Delta R_{12}(\Delta\eta; x_0, y_0)$ needs to be compensated. However, the phase related to $\Delta R_{12}(\Delta\eta; x_0, y_0)$ varies with azimuth time and clutter's position, and signals of clutter scattering points are aliased in echo domain. Therefore, it is difficult to directly compensate the phase in echo domain. Additionally, the differences of ΔR_{12} and $\Delta\Phi_{12}$ are caused by the movement of the transmitting platform. Therefore, if the influence of the transmitter's motion can be eliminated, the phase compensation processing will be simplified.

From Eqs. (4.69) and (4.71) in the previous section, it can be find that the bistatic range history of stationary clutter includes a constant term, the first-order term of azimuth time and the second-order term of azimuth time. Thus, in order to carry out the phase compensation effectively, these terms related to azimuth time in stationary clutter range history can be eliminated first. Then, the range histories of the stationary clutter of two channels are

$$R_1(\eta_1, x_0, y_0) = R_{T,c} + R_{R1,c} \tag{4.75}$$

$$R_2(\eta_1, x_0, y_0) = R_{T,c} + R_{R2,c} \tag{4.76}$$

At this moment, there is only a constant term in range history, and it is merely related the spatial position of the stationary clutter. Thus, the phase difference is only related to the constant term, given by

$$\Delta\Phi_{12}(\Delta\eta; x_0, y_0) = 2\pi \frac{R_{R1,c} - R_{R2,c}}{\lambda} \tag{4.77}$$

Equation (4.77) represents a constant phase, while $\Delta\Phi_{12}(\Delta\eta; x_0, y_0)$ is still spatial variant. To realize different phase compensation processing for different stationary clutter scattering point at different position, clutter points need to be focused on its own position before phase compensation.

Clutter signal in channel 1 after range compression can be expressed as:

$$s_{1,s}(\tau, \eta; x_0, y_0) = \sigma(x_0, y_0) \cdot |K_r|T_p \, \mathrm{sinc}\left\{\pi K_r T_p\left[\tau - \frac{R_{1,s}(\eta; x_0, y_0)}{c}\right]\right\}$$
$$\cdot \exp\left\{-j2\pi f_0 \frac{R_{1,s}(\eta; x_0, y_0)}{c}\right\} \tag{4.78}$$

where

$$R_{1,s}(\eta, x_0, y_0) \approx R_{T,c} + R_{R1,c} + (-V_T \sin\theta_T - V_R \sin\theta_{R1})\eta$$
$$+ \frac{1}{2}\left(\frac{V_T^2 \cos^2\theta_T}{R_{T,c}} + \frac{V_R^2 \cos^2\theta_{R1}}{R_{R1,c}}\right)\eta^2 \tag{4.79}$$

Clutter signal in channel 2 after range compression can be expressed as:

$$s_{2,s}(\tau, \eta; x_0, y_0) = \sigma(x_0, y_0) \cdot |K_r|T_p \, \mathrm{sinc}\left\{\pi K_r T_p\left[\tau - \frac{R_{2,s}(\eta; x_0, y_0)}{c}\right]\right\}$$
$$\cdot \exp\left\{-j2\pi f_0 \frac{R_{2,s}(\eta; x_0, y_0)}{c}\right\} \tag{4.80}$$

where

$$R_{2,s}(\eta, x_0, y_0) \approx R_{T,c} + R_{R2,c} + (-V_T \sin\theta_T - V_R \sin\theta_{R2})\eta$$
$$+ \frac{1}{2}\left(\frac{V_T^2}{R_{T,c}} \cos^2\theta_T + \frac{V_R^2}{R_{R2,c}} \cos^2\theta_{R2}\right)\eta^2 \tag{4.81}$$

According to the analysis in previous chapter, range cell migration exists in clutter echo in bistatic SAR. Thus, range cell migration correction is necessary before phase compensation. To eliminate the terms related to azimuth time in range history, the compensation functions of two receiving channels can be constructed respectively:

$$C_1(\eta) = \exp\left\{\frac{j2\pi}{\lambda}\left[\begin{array}{c}(-V_T \sin\theta_T - V_R \sin\theta_{R1})\eta \\ +\frac{1}{2}\left(\frac{V_T^2 \cos^2\theta_T}{R_{T,c}} + \frac{V_R^2 \cos^2\theta_{R1}}{R_{R1,c}}\right)\eta^2\end{array}\right]\right\} \quad (4.82)$$

$$C_2(\eta) = \exp\left\{\frac{j2\pi}{\lambda}\left[\begin{array}{c}(-V_T \sin\theta_T - V_R \sin\theta_{R2})\eta \\ +\frac{1}{2}\left(\frac{V_T^2 \cos^2\theta_T}{R_{T,c}} + \frac{V_R^2 \cos^2\theta_{R2}}{R_{R2,c}}\right)\eta^2\end{array}\right]\right\} \quad (4.83)$$

After range cell migration correction and multiplying the compensation functions with the echo signals in Eqs. (4.78) and (4.80), we have

$$S_{1,s}(\tau, \eta) = s_{1,s}(\tau, \eta; x_0, y_0) \cdot C_1(\eta) = D_1(\tau)\exp\left\{-j\frac{2\pi}{\lambda}(R_{T,c} + R_{R1,c})\right\} \quad (4.84)$$

$$S_{2,s}(\tau, \eta) = s_{2,s}(\tau, \eta; x_0, y_0) \cdot C_2(\eta) = D_2(\tau)\exp\left\{-j\frac{2\pi}{\lambda}(R_{T,c} + R_{R2,c})\right\} \quad (4.85)$$

where

$$D_1(\tau) = \sigma(x_0, y_0) \cdot |K_r|T_p \sin c\left\{\pi K_r T_p\left[\tau - \frac{R_{1,s}(\eta; x_0, y_0)}{c}\right]\right\} \quad (4.86)$$

$$D_2(\tau) = \sigma(x_0, y_0) \cdot |K_r|T_p \sin c\left\{\pi K_r T_p\left[\tau - \frac{R_{2,s}(\eta; x_0, y_0)}{c}\right]\right\} \quad (4.87)$$

After phase compensation, the remaining clutter range history only contains a constant term depended on the clutter point's position, as shown in Eq. (4.84) and (4.85). Then, transforming the echo signals of two receiving channels into Doppler domain by FFT, we have

$$I_1(\tau, f_\eta) = FFT(S_{1,s}(\tau, \eta)) = D_1(\tau) \cdot \exp\left\{-j\frac{2\pi}{\lambda}(R_{T,c} + R_{R_1,c})\right\} \cdot H \quad (4.88)$$

$$I_2(\tau, f_\eta) = FFT(S_{2,s}(\tau, \eta)) = D_2(\tau) \cdot \exp\left\{-j\frac{2\pi}{\lambda}(R_{T,c} + R_{R2,c})\right\} \cdot H \quad (4.89)$$

where

$$H = \int_0^{T_p} \exp\{-j2\pi f_\eta \eta\}d\eta \quad (4.90)$$

From Eqs. (4.88) and (4.89), only the constant terms in range history are different. Then, the phase compensation on channel 1 is operated to eliminate the difference between Eqs. (4.88) and (4.89). In addition, since clutter echo in two channels are focused after FFT, the spatial variation problem of the constant phase can be solved by calculating the corresponding phase of each pixel position. The compensated phase can be obtained as

$$G(x_0, y_0) = \exp\left\{-j\frac{2\pi}{\lambda}\left(R_{R2,c} - R_{R1,c}\right)\right\} \tag{4.91}$$

Then, $I_1'\left(\tau, f_\eta\right) = I_1\left(\tau, f_\eta\right) \cdot G(x_0, y_0) = D_1(\tau) \cdot \exp\left\{-j\frac{2\pi}{\lambda}\left(R_{T,c} + R_{R_2,c}\right)\right\} \cdot H$.

It can be observed that clutter echo in two channels are the same. After echo cancellation of two receiving channels, the following equation can be obtained

$$I_{12}(\tau, f_a) = I_1'\left(\tau, f_\eta\right) - I_2\left(\tau, f_\eta\right) = 0 \tag{4.92}$$

Thus, bistatic SAR clutter has been suppressed. While, due to the motion of target, the signals in two channels are not the same after focusing and phase compensation. Thus, the information of moving target can be retained in RD-based DPCA clutter suppression method.

4.2.1.3 Simulation Verification

The simulation is based on the geometrical parameters given in Table 4.1 and signal parameters given in Table 4.2.

(1) Set two moving targets, with the same original position $p_m = (0, 0)m$. The velocities of moving target 1 are $v_x = 2$ m/s and $v_y = 1$ m/s, while the velocities of moving target 2 are $v_x = 3$ m/s and $v_y = 2$ m/s. Let the stationary clutter scattering point locate at the center of the scene, with position $p_s = (0, 0)m$. Thus, two moving

Table 4.1 Geometrical configuration parameters of RD-DPCA simulation

Parameters	Values
Transmitter location	[−10000 m, −13000 m, 12000 m]
Receiver location	[0 m, −10000 m, 10000 m]
Transmitter velocity	200 m/s
Receiver velocity	200 m/s
Channel space	0.4 m

Table 4.2 Signal parameters of RD-DPCA simulation

Parameters	Values
Pulse width	1.5 us
Band width	150 MHz
Carrier frequency	9.6 GHz
Synthetic aperture time	1 s
PRF	1000 Hz

targets and clutter point are located at the same place at the initial moment. The simulation results are shown in Fig. 4.14.

From Fig. 4.14a, moving target after imaging in channel 1 produces a position shift in the azimuth direction. The position shift occurs in two channels and target 2 moves farther as it has a larger velocity. Besides, images of moving targets are slightly defocused. Comparing Fig. 4.14a with Fig. 4.14b, it can be found that imaging positions of both targets in two channels are the same. From Fig. 4.14c, after two-channel cancellation, the signal of stationary clutter point has been eliminated, while the two moving targets are well retained in image domain.

$A_{m,e}$ and $A_{c,e}$ represent maximum amplitudes of moving target signal and stationary clutter signal in echo domain, respectively. According to the formula $SCR_i = 20 \cdot \log_{10}(A_{m,e}/A_{c,e})$, the input SCR can be expressed as: $SCR_i = 6.02$ dB. $A_{m,i}$ and $A_{c,i}$ represent maximum amplitudes of the remaining moving target signal

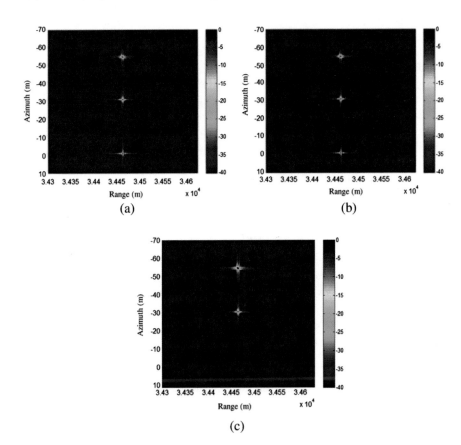

Fig. 4.14 Clutter suppression results based on RD-based DPCA. **a** Image result obtained in channel 1; **b** Image result obtained in channel 2; **c** Cancellation result

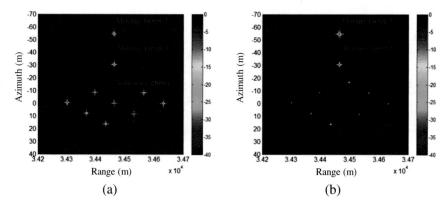

Fig. 4.15 Clutter suppression results with multiple clutter scattering points. **a** Image result obtained in channel 1; **b** Cancellation result

and the remaining stationary clutter signal respectively in image domain after two-channel cancellation. According to formula $SCR_o = 20 \cdot \log_{10}(A_{m,i}/A_{c,i})$, the output SCR can be expressed as $SCR_o = 47.78$ dB after clutter cancellation.

From Fig. 4.15, it can be seen that this method also has different suppression effect on different stationary clutter points at different locations. Similarly, the closer the stationary clutter is to the scene center, the more effective clutter suppression effect is. Therefore, for this method, the processing scene area should not to be too large, otherwise, the remaining clutter energy will have an impact on the following detection performance.

4.2.2   BP-Based DPCA Method

In order to solve the problem that two-channel equivalent phase centers are not coincided in bistatic SAR, this part will utilize Back Projection (BP) algorithm [81] to compensate the range history of stationary clutter, through the time delay phase compensation on every imaging grid in the whole synthetic aperture time. And then, suppress bistatic SAR clutter will be suppressed in image domain by two-channel cancellation.

The configuration of bistatic SAR with two receiving channels is shown in Fig. 4.16. The coordinate of the transmitting channel is (x_T, y_T, h_T), and the velocity of the transmitter is $(0, V, 0)$. There are two receiving channels located along-track on the receiving platform. The distance between two channels is d. The coordinate of the receiving channel 2 is (x_R, y_R, h_R), and the coordinate of the receiving channel 1 is $(x_R, y_R - d, h_R)$. The velocity of the receiver is $(0, V, 0)$.

Let the velocity of moving target $P(x, y, 0)$ be $v_r = (v_x, v_y)$, whose components along x and y axes are v_x and v_y, respectively. Echo signals after range compression in channel 1 and channel 2 can be expressed as

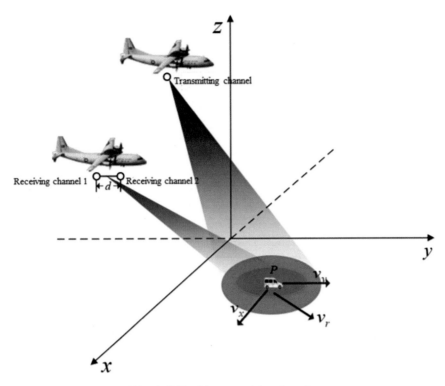

Fig. 4.16 Configuration of bistatic SAR with two receiving channels

$$S_{r1}(\tau, \eta) = \sigma \, sinc\left\{\pi K_r T_p\left[\tau - \frac{R_1^*(\eta)}{c}\right]\right\} \exp\left\{-j2\pi f_0 \frac{R_1^*(\eta)}{c}\right\} \tag{4.93}$$

$$S_{r2}(\tau, \eta) = \sigma \, sinc\left\{\pi K_r T_p\left[\tau - \frac{R_2^*(\eta)}{c}\right]\right\} \exp\left\{-j2\pi f_0 \frac{R_2^*(\eta)}{c}\right\} \tag{4.94}$$

where τ is the range fast time, η is the azimuth slow time, σ is the target amplitude, K_r is frequency rate of the transmitting signal, T_p is the pulse width, f_0 is the carrier frequency, and c is the propagating velocity of microwaves.

$$\begin{cases} R_1^*(\eta) = R_T^*(\eta) + R_{R1}^*(\eta) \\ R_2^*(\eta) = R_T^*(\eta) + R_{R2}^*(\eta) \end{cases} \tag{4.95}$$

$$
\begin{cases}
R_T^*(\eta) = \sqrt{\begin{array}{c}(x_T - x - v_x \cdot \eta)^2 + \\ (y_T + V \cdot \eta - y - v_y \cdot \eta)^2 + h_T^2\end{array}} \\[3mm]
R_{R1}^*(\eta) = \sqrt{\begin{array}{c}(x_R - x - v_x \cdot \eta)^2 + \\ (y_R - d + V \cdot \eta - y - v_y \cdot \eta)^2 + h_R^2\end{array}} \\[3mm]
R_{R2}^*(\eta) = \sqrt{\begin{array}{c}(x_R - x - v_x \cdot \eta)^2 + \\ (y_R + V \cdot \eta - y - v_y \cdot \eta)^2 + h_R^2\end{array}}
\end{cases}
\tag{4.96}
$$

After Taylor series expansion on the bistatic range history at the beam center crossing time $\eta = 0$, the echo signals in two channels can be expressed as:

$$
\begin{cases}
S_{r1}(\tau, \eta) \approx D_1(\tau) \exp\left\{ -j \dfrac{2\pi}{\lambda} \left(r_{1,0} + r_{1,1} \cdot \eta + r_{1,2} \cdot \eta^2 \right) \right\} \\[2mm]
r_{1,0} = R_{T,0} + R_{R1,0} \\[2mm]
r_{1,1} = (-V \sin \theta_T - V \sin \theta_{R1} + v_{rT} + v_{rR}) \\[2mm]
r_{1,2} = \dfrac{V_{v_y}^2 + v_x^2}{2 R_{R1,0}} - \dfrac{\left(v_x(x - X_r) - V_{v_y}(y - Y_r) \right)^2}{2 R_{R1,0}^3} + \dfrac{V_{v_y}^2 + v_x^2}{2 R_{T,0}} - \dfrac{\left(x v_x - y V_{v_y} \right)^2}{2 R_{T,0}^3}
\end{cases}
\tag{4.97}
$$

$$
\begin{cases}
S_{r2}(\tau, \eta) \approx D_2(\tau) \exp\left\{ -j \dfrac{2\pi}{\lambda} \left(r_{2,0} + r_{2,1} \cdot \eta \right) \right\} \\[2mm]
r_{2,0} = R_{T,0} + R_{R2,0} \\[2mm]
r_{2,1} = (-V \sin \theta_T - V \sin \theta_{R2} + v_{rT} + v_{rR}) \\[2mm]
r_{2,2} = \dfrac{V_{v_y}^2 + v_x^2}{2 R_{R2,0}} - \dfrac{\left(v_x(x - X_r) - V_{v_y}(y - Y_r) \right)^2}{2 R_{R2,0}^3} + \dfrac{V_{v_y}^2 + v_x^2}{2 R_{T,0}} - \dfrac{\left(x v_x - y V_{v_y} \right)^2}{2 R_{T,0}^3}
\end{cases}
\tag{4.98}
$$

$$
\begin{cases}
R_{T,0} = \sqrt{(x_T - x)^2 + (y_T - y)^2 + h_T^2} \\[2mm]
R_{R1,0} = \sqrt{(x_R - x)^2 + (y_R - d - y)^2 + h_R^2} \\[2mm]
R_{R2,0} = \sqrt{(x_R - x)^2 + (y_R - y)^2 + h_R^2} \\[2mm]
V_{v_y} = V - v_y
\end{cases}
\tag{4.99}
$$

$$
\begin{cases}
D_1(\tau) = \sigma \operatorname{sinc}\left\{ \pi K_r T_p \left[\tau - \dfrac{R_1^*(\eta)}{c} \right] \right\} \\[2mm]
D_2(\tau) = \sigma \operatorname{sinc}\left\{ \pi K_r T_p \left[\tau - \dfrac{R_2^*(\eta)}{c} \right] \right\}
\end{cases}
\tag{4.100}
$$

where v_{rT} and v_{rR} are radial velocities of moving target relative to the transmitter and the receiver.

While choosing the size of the scene and the length of the sampling interval, the time from the radar position to every pixel in the imaging area of each channel can be expressed as:

$$t_{ij1} = \frac{R'_T(\eta) + R'_{R1}(\eta)}{c}, \quad t_{ij2} = \frac{R'_T(\eta) + R'_{R2}(\eta)}{c} \tag{4.101}$$

where $R'_T(\eta)$, $R'_{R1}(\eta)$ and $R'_{R2}(\eta)$ are distances between BP grid point (x', y') relative to the transmitter, the receiving channel 1 and channel 2 at azimuth moment η.

After interpolation on echoes from channel 1 followed with compensation phase factor $\exp\{j2\pi f_c t_{ij1}\}$ and channel 2 followed with compensation phase factor $\exp\{j2\pi f_c t_{ij2}\}$, we have

$$\begin{cases} S_1(\tau, \eta) = D_1(\tau) \cdot F_1(\eta) \cdot \exp\left\{-j\frac{2\pi}{\lambda}\left(R_{T,0} - R'_{T,0} + R_{R1,0} - R'_{R1,0}\right)\right\} \\ \\ S_2(\tau, \eta) = D_2(\tau) \cdot F_2(\eta) \cdot \exp\left\{-j\frac{2\pi}{\lambda}\left(R_{T,0} - R'_{T,0} + R_{R2,0} - R'_{R2,0}\right)\right\} \end{cases}$$
$$\tag{4.102}$$

where $R'_{T,0}$, $R'_{R1,0}$ and $R'_{R2,0}$ represent Taylor series expansion coefficients of constant terms in $R'_T(\eta)$, $R'_{R1}(\eta)$ and $R'_{R2}(\eta)$ respectively.

$$\begin{cases} F_1(\eta) = \exp\left\{-j\frac{2\pi}{\lambda}\left[\begin{array}{c} r_{1,1}\eta + V\left(\sin\theta'_T + \sin\theta'_{R1}\right)\eta + \\ r_{1,2}\eta^2 - \frac{1}{2}\left(\frac{V^2\cos^2\theta'_T}{R'_{T,0}} + \frac{V^2\cos^2\theta'_{R1}}{R'_{R1,0}}\right)\eta^2 \end{array}\right]\right\} \\ \\ F_2(\eta) = \exp\left\{-j\frac{2\pi}{\lambda}\left[\begin{array}{c} r_{2,1}\eta + V\left(\sin\theta'_T + \sin\theta'_{R2}\right)\eta + \\ r_{2,2}\eta^2 - \frac{1}{2}\left(\frac{V^2\cos^2\theta'_T}{R'_{T,0}} + \frac{V^2\cos^2\theta'_{R2}}{R'_{R2,0}}\right)\eta^2 \end{array}\right]\right\} \end{cases}$$
$$\tag{4.103}$$

$$\begin{cases} \theta_T = \arcsin\left(\frac{y - y_T}{R_{T,0}}\right) \\ \theta_{R1} = \arcsin\left[\frac{y - (y_R - d)}{R_{R1,0}}\right] \\ \theta_{R2} = \arcsin\left(\frac{y - y_R}{R_{R2,0}}\right) \end{cases} \begin{cases} \theta'_T = \arcsin\left(\frac{y' - y_T}{R'_{T,0}}\right) \\ \theta'_{R1} = \arcsin\left[\frac{y' - (y_R - d)}{R'_{R1,0}}\right] \\ \theta'_{R2} = \arcsin\left(\frac{y' - y_R}{R'_{R2,0}}\right) \end{cases} \tag{4.104}$$

Since the distance between two receiving channels d is very small, an assumption $F(\eta) = F_1(\eta) \approx F_2(\eta)$ can be made. After coherent integration, the imaging result can be expressed as:

$$f_1(x_i, y_j) = \int_\eta S_1(\tau, \eta)d\eta$$

$$= \int_\eta D_1(\tau)F(\eta)d\eta \cdot \exp\left\{-j\frac{2\pi}{\lambda}\left(R_{T,0} - R'_{T,0} + R_{R1,0} - R'_{R1,0}\right)\right\}$$

$$= \rho(x_i, y_j) \cdot \exp\left\{-j\frac{2\pi}{\lambda}\left(R_{T,0} - R'_{T,0} + R_{R1,0} - R'_{R1,0}\right)\right\} \qquad (4.105)$$

$$f_2(x_i, y_j) = \int_\eta S_2(\tau, \eta)d\eta$$

$$= \int_\eta D_2(\tau)F(\eta)d\eta \cdot \exp\left\{-j\frac{2\pi}{\lambda}\left(R_{T,0} - R'_{T,0} + R_{R2,0} - R'_{R2,0}\right)\right\}$$

$$= \rho(x_i, y_j) \cdot \exp\left\{-j\frac{2\pi}{\lambda}\left(R_{T,0} - R'_{T,0} + R_{R2,0} - R'_{R2,0}\right)\right\} \qquad (4.106)$$

where $\rho(x_i, y_j)$ is the image amplitude after BP algorithm.

Then, subtracting $f_1(x_i, y_j)$ with $f_2(x_i, y_j)$, the cancellation result is given by

$$f_{12}(x_i, y_j) = f_1(x_i, y_j) - f_2(x_i, y_j)$$

$$= f_2(x_i, y_j)\left(\frac{f_1(x_i, y_j)}{f_2(x_i, y_j)} - 1\right)$$

$$= f_2(x_i, y_j)\left(\exp\left\{-j\frac{2\pi}{\lambda}\left[(R_{R1,0} - R_{R2,0}) - (R'_{R1,0} - R'_{R2,0})\right]\right\} - 1\right)$$

$$\qquad (4.107)$$

After clutter cancellation, its phase can be expressed as

$$\varphi_{BP} = -\left\{\frac{2\pi}{\lambda}\left[(R_{R1,0} - R_{R2,0}) - (R'_{R1,0} - R'_{R2,0})\right]\right\} \qquad (4.108)$$

For the stationary clutter, the azimuth Doppler phase can be well compensated, that is, $R_{R,0} = R'_{R,0}$, $R_{R1,0} = R'_{R1,0}$, $R_{R2,0} = R'_{R2,0}$. Therefore, from Eqs. (4.107) and (4.108), the stationary clutter can be completely suppressed after clutter cancellation. Due to the target position shift on the BP grid for moving target, we can obtain: $R_{T,0} \neq R'_{T,0}$, $R_{R1,0}R'_{R1,0}$, $R_{R2,0} \neq R'_{R2,0}$. Thus, from Eqs. (4.107) and (4.108), moving target is retained after cancellation, with non-zero phase.

The following part is the BP-DPCA simulation with parameters in Tables 4.3 and 4.4.

(1) Point target simulation

Set two moving targets and one stationary clutter point with the same original position (0, 0, 0) m, where the velocities of target 1 are $v_x = 1m/s$, $v_y = 1m/s$ and the velocities of target 2 are $v_x = 2m/s$ and $v_y = 2m/s$.

Parameters	Values
Channel space	0.2 m
Transmitter location	$(-10, -12, 10)$ km
Receiver location	$(0, -8, 8)$ km
Transmitter velocity	200 m/s
Receiver velocity	200 m/s

Table 4.3 Geometrical configuration parameters of BP-DPCA simulation

Parameters	Values
Bandwidth	150 MHz
Carrier frequency	10 GHz
PRF	2000 Hz

Table 4.4 Signal parameters of BP-DPCA simulation

As shown in Fig. 4.17a, when the target is moving, there will be an obvious position shift in the image during BP processing, and the position shift of two moving targets are different, for greater velocity with larger shift. From Fig. 4.17b, after BP–DPCA clutter suppression in image domain, the stationary clutter point has been eliminated, while the moving target is retained, for greater speed with more moving target remaining.

Area target simulation

In the background of area target, set one moving target with the original position $(0, 0, 0)m$ and velocities of $v_x = 2m/s$, $v_y = 2m/s$.

From Fig. 4.18, before clutter cancellation, moving target is difficult to be distinguished from the stationary clutter background. After BP-DPCA clutter cancellation

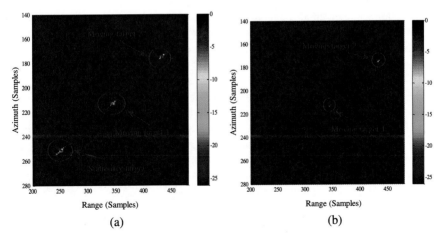

(a)   (b)

Fig. 4.17 Clutter suppression results of BP-DPCA in image domain. **a** Before clutter cancellation; **b** After clutter cancellation

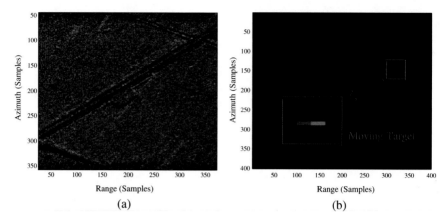

Fig. 4.18 BP-DPCA clutter suppression result with area clutter background. **a** Before clutter cancellation; **b** After clutter cancellation

in image domain, the stationary clutter background has been effectively cancelled and moving target is retained.

4.3   Summary

Two kinds of DPCA-based clutter suppression methods are presented in this chapter to achieve bistatic SAR clutter suppression. One is the multi-pulse DPCA clutter suppression method, which performs RCMC for each receiving channels to compensate the range cell migration, and uses Doppler parameter equalization method to eliminate the space–time spectrum extension of bistatic SAR clutter. Then the clutter suppression is realized through the multi-pulse two-channel cancellation. Since clutter non-stationarity does not affect this method, this method has better clutter suppression effect than traditional STAP method in bistatic SAR. The other kind of method is image-domian DPCA, including RD-based DPCA and BP-based DPCA. In RD-based DPCA method, the influence of transmitter motion on the equivalent phase center can be eliminated by compensating terms related to stationary clutter range history and azimuth time. Due to the velocity of moving target, the information of the moving target can be retained after RD-base DPCA. In the BP-based DPCA method, the range history of stationary clutter is compensated by the BP algorithm in image domain, and the clutter can be suppressed by two-channel cancellation. The simulation results have porved that DPCA-based clutter suppression methods mentioned above can effectively suppress bistatic SAR clutter and retain moving target's information.

Chapter 5
Optimization-Based Clutter Suppression Method

Abstract This chapter introduces an optimization-based clutter suppression method for bistatic SAR clutter suppression. The main idea of the proposed method is to design and generate a suppression filter in space–time domain, whose space–time frequency response is matched with clutter spectrum. Based on the clutter modelling in the previous chapter, space–time information of clutter spectrum can be obtained first. And then, the suppression filter can be designed and the filter weight calculation process is transferred into a constrained optimization problem (COP), according to the obtained clutter space–time information. Finally, the particle swarm optimization (PSO) algorithm is applied to solve COP and obtain the optimal solution, i.e., the desired matched space–time filter weight, for BiSAR non-stationary clutter suppression. Since the generation of the designed filter circumvents CCM estimation, the proposed method will not be affected by the non-stationary characteristic of BiSAR clutter.

Keywords Space–time filter design · Constrained optimization problem · Particle swarm optimization · Non-stationary clutter suppression · Matched space–time filter

5.1 Space–Time Filter Design and Optimally Modelling

The key of clutter suppression is to construct a suitable filter to distinguish target and undesired clutter. In STAP methods, the suppression filter is set up by minimizing the output power of clutter echo. According to the linear constraint minimum variance (LCMV) criterion, the suppression filter weight can be obtained by the inversion of CCM and target's steering vector [44–47]. However, its performance heavily depends on the accuracy of the estimated CCM. Based on the previous analysis, the strong non-stationarity of BiSAR clutter will lead to an inaccurate CCM in traditional STAP [57, 79]. Consequently, a worse or even unacceptable clutter suppression would be resulted in.

Therefore, to effectively suppress the non-stationary clutter in BiSAR, an optimization-based clutter suppression method is proposed, which is called optimally matched space–time filtering (MSTF). MSTF method generates the suppression filter

© The Author(s), under exclusive license to Springer Nature Singapore Pte Ltd. 2022   79
Z. Li et al., *Bistatic SAR Clutter Suppression*,
https://doi.org/10.1007/978-981-19-0159-1_5

to circumvent the CCM estimation and the space–time response (2D responese) of the filter is designed to be matched with clutter spectrum. Thus, clutter suppression with the proposed method does not rely on the CCM estimation and sufficient IID secondary samples. In this section, the matched space–time filter is designed first and then its filter weight calculation problem is modelled as an optimization problem with a constraint.

According to the analysis in Chap. 3, clutter ridges with different bistaitc ranges are different in space–time domain and they are related to BiSAR configuration [57]. while for a particular bistatic range, the space–time characteristic of clutter is fixed. Thus, we can construct a space–time filter whose notch is located at the corresponding space–time position, where clutter is located at, to match clutter spectrum. Then, BiSAR nonstationary clutter suppression can be realized by the designed filter.

To design the desired filter, four factors are considered:

(1)   the space–time position of the space–time filter notch;
(2)   the depth of the space–time filter notch;
(3)   the width of the space–time filter notch;
(4)   the filter response of moving target of the space–time filter.

The first consideration is to adjust the notch position of the space–time filter so that it can be coincided with clutter ridge in space–time domain. Thus, BiSAR clutter signal can be filtered accordingly. The second consideration is to make the notch of the designed space–time filter deep enough, so as to ensure that the BiSAR clutter can be fully suppressed. The third consideration is to make the notch width of the designed space–time filter matched with the real clutter spectrum, so as to ensure that the filter notch is not broadened under BiSAR configuration, and improve the performance of clutter filtering and detection of target with slow velocity. The fourth consideration is to make moving target signal retained as much as possible after clutter suppression. Figure 5.1 shows the diagram of the space–time filter design.

Assume the receiver of BiSAR has N receiving channels and K pulses are processed in one coherent processing interval (CPI). Therefore, the designed filter weight W_d can be expressed by an $NK \times 1$ dimension complex vector:

$$W_d = [W_{d1}, W_{d2}, ..., W_{dNK}]^T \tag{5.1}$$

Then, two-dimensional frequency response of the filter in space–time domain can be calculated by the following formula:

$$R_{f_d, f_s}(W_d) = W_d^H \Lambda(f_d, f_s) \tag{5.2}$$

where $\Lambda(f_d, f_s) = S_s(f_s) \otimes S_t(f_d) \in NK \times 1$ is the space–time steering vector. f_d and f_s are the normalized Doppler frequency and spatial frequency, respectively.

Similarly, the two-dimensional frequency response of the filter to moving target can be expressed as

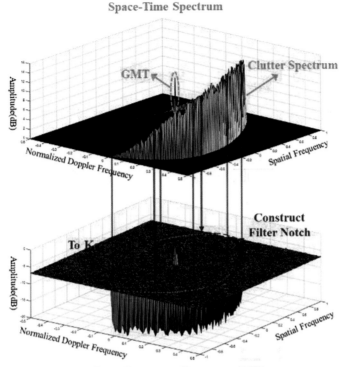

Fig. 5.1   The diagram of the space–time filter design

$$R_{GMT}(W_d) = W_d^H \cdot (V_s(f_{sMT}) \otimes V_t(f_{dMT})) \tag{5.3}$$

where $V_s(f_{sMT}) \in N \times 1$ and $V_t(f_{dMT}) \in K \times 1$ are the space and time steering vectors of moving target, respectively.

In the following, the space–time filter will be designed and the weight acquisition can be optimally modelled as a constrained optimization problem (COP). For the BiSAR system, the background echo can be regarded as the sum of echo of plentiful scattering points. The number of these scattering points on the ground that can be distinguished during a synthetic aperture time that is usually determined by the azimuth resolution of system. Thus, the number of scattering point in the aperture can be expressed as

$$Q = L_{syn}/\rho_a \tag{5.4}$$

where $L_{syn} = V_R T_{syn}$ is the length of the synthetic aperture and T_{syn} is the synthetic aperture time. The azimuth resolution ρ_a can be expressed as

$$\rho_a = \frac{k_1 \lambda}{T_{syn} \| \omega_{TA} + \omega_{RA} \|} \tag{5.5}$$

where k_1 is a constant. ω_{TA} and ω_{RA} are angular velocities of the transmitter and receiver, respectively.

Based on the above analysis, clutter echo received by the BiSAR system can be regarded as the sum of echo reflected by the clutter scattering points in Q directions. That is to say, BiSAR clutter in the certain range cell is considered to be distributed at Q positions in space–time domain. Thus, the suppression filter notch should be set at the corresponding Q space–time positions. Dividing the receiving beamwidth uniformly into Q directions, the two-dimensional frequency response of the filter can be expressed as

$$F(W_d) = \begin{bmatrix} R_{f_{d1}, f_{s1}}(W_d) \\ R_{f_{d2}, f_{s2}}(W_d) \\ \vdots \\ R_{f_{dQ}, f_{sQ}}(W_d) \end{bmatrix} = \begin{bmatrix} W_d^H(S_s(f_{s1}) \otimes S_t(f_{d1})) \\ W_d^H(S_s(f_{s2}) \otimes S_t(f_{d2})) \\ \vdots \\ W_d^H(S_s(f_{sQ}) \otimes S_t(f_{dQ})) \end{bmatrix} \tag{5.6}$$

where $f_{dp}, p = 1, 2, \cdots Q$ and $f_{sp}, p = 1, 2, \cdots Q$ are the normalized Doppler frequency and spatial frequency of clutter in the $p-th$ direction.

According to the Eq. (5.6), the space–time position of the filter notch is confirmed. In order to suppress BiSAR clutter effectively, $F(W_d^*)$ should be set as zero, where $W_d^* = \begin{bmatrix} W_{d1}^*, W_{d2}^*, ..., W_{dNK}^* \end{bmatrix}^T \in D$ is the decision vector composed with NK complex decision variables, and D is the decision space. It can be seen that when $F(W_d^*) = \mathbf{0}$ is satisfied, clutter suppression can be realized and the solution W_d^* is the desired filter weight.

Generally, Newton-like method [82, 83] is the most classical method for solving equations. In Newton-like method, the gradient information is used to find the fastest descending direction of the equation, and the Jacobian matrix or difference matrix is constructed as the iterative matrix. By iteratively updating the iteration matrix and the current solution, the optimal solution can be gradually approached. The iterative formula in Newton method is given as follows

$$x(g + 1) = x(g) - M^{-1}(x(g), v(g))F(x(g)) \tag{5.7}$$

where $x(g)$ is the solution after the $g - th$ iteration. $M(x(g), v(g))$ is the difference coefficient matrix of equations $F(x(g))$ given by

$$M(x(g), v(g)) = \begin{bmatrix} \frac{[F(x(g)+v_1(g) \cdot e_1) - F(x(g))]}{v_1(g)} \\ \vdots \\ \frac{[F(x(g)+v_Q(g) \cdot e_Q) - F(x(g))]}{v_Q(g)} \end{bmatrix}^T \tag{5.8}$$

where $v_i(g)=\mu_i\|F(x(g))\|$, $i = 1, 2, \ldots, Q$ is step factor and μ_i is a constant which cannot be zero. e_i is the vector, whose $i-th$ component is 1 and other components are 0.

Although Newton-like method has advantages of fast convergence and high accuracy in solving equations, it cannot be applied to solve $F(W_d^*) = \mathbf{0}$ for filter weight calculation in BiSAR clutter suppression. The main reasons are given as follows. Firstly, the convergence of Newton-like method depends on its iterative matrix. However, in the equation $F(W_d^*) = \mathbf{0}$, the dimension of W_d^* is $NK \times 1$ and Q is usually different from system DOF ($Q \neq NK$). The iteration matrix obtained in Eq. (5.8) will not be a square matrix. If the pseudo inverse matrix is used to replace the inverse of the iterative matrix, the iterative updating process will be unstable and the method may not converge. Secondly, the prerequisite for fast convergence of Newton-like method is that the initial solution is close enough to the true solution. In practice, it is difficult to know the information of the true solution before solving the equations. Thus, the convergence of Newton-like method cannot be obtained. Thirdly, since the gradient information used by Newton-like method is the local information near the current solution $x(g)$, this kind of method is easy to fall into the local optimization, resulting in the failure of obtaining the global optimal solution.

In order to construct the space–time matched filter for BiSAR clutter suppression, the problem of calculating the filter weight is modelled as a constrained optimization problem, and then the swarm intelligence optimization algorithm is applied to solve the problem mentioned above. Thus, the optimal filter weight can be obtained without gradient information and a better convergence performance can be also gained. According to the filter's frequency response of clutter and moving target, the constrained optimization problem is constructed as follows

$$
\begin{cases}
\min F_0(W_d) = \min \sum_{i=1}^{3} f_i(W_d) \\
s.t. \quad |R_{GMT}(W_d) - 1| \leq \varepsilon
\end{cases}
\tag{5.9}
$$

where ε is the tolerance of noise error. The constraint condition is applied to make the designed filter response of GMT close to 1 in space–time domain, which means that target signal can be totally retained after space–time filtering. $\{f_i(W_d)|i = 1, 2, 3\}$ is the constructed objective function, which aims at limiting the depth and width of the filter notch. The objective function $\{f_i(W_d)|i = 1, 2, 3\}$ is given by

$$
f_1(W_d) = \frac{1}{Q} \sum_{i=1}^{Q} \frac{W_d^H(S_s(f_{si}) \otimes S_t(f_{di}))}{R_{GMT}(W_d)}
\tag{5.10}
$$

$$
f_2(W_d) = \frac{1}{Q} \sum_{i=1}^{Q} \left(\frac{W_d^H(S_s(f_{si}) \otimes S_t(f_{di}))}{R_{GMT}(W_d)} - f_1(W_d) \right)^2
\tag{5.11}
$$

$$f_3(W_d) = -\frac{1}{2Q}\frac{(sum\,R_L(W_d) + sum\,R_R(W_d))}{R_{GMT}(W_d)} \tag{5.12}$$

In the above equations, the objective function $f_1(W_d)$ and $f_2(W_d)$ will jointly determine the depth of filter notch. $f_1(W_d)$ is the average value of the filter notch's response and $f_2(W_d)$ is the variance of the filter notch's response. The smaller the average value of the amplitude response is (5.10), the deeper depth of the filter notch can be generated at the corresponding Q positions in space–time domain. The variance in (5.11) is utilized to limit the fluctuation of the notch depth. Thus, by minimizing $f_1(W_d)$ and $f_2(W_d)$, the filter notch is restricted to be deep and smooth. $f_3(W_d)$ in (5.12) is the average value of the 2-D response around the notch and it's applied to limit the width of the filter notch. From (5.12), it can be seen that $f_3(W_d)$ is contributed by $sum\,R_L(W_d)$ and $sum\,R_R(W_d)$. The first term in $f_3(W_d)$ is the sum response of the left side of filter notch, and the second term is that of the right side. The width of the filter notch is determined by a constant Doppler frequency Δf_d. Two terms $sum\,R_L(W_d)$ and $sum\,R_R(W_d)$ can be expressed as

$$sum\,R_L(W_d) = \sum_{j=1}^{Q} W_d^H \left(S_s(f_{sj}) \otimes S_t(f_{dj} - \Delta f_d) \right) \tag{5.13}$$

$$sum\,R_R(W_d) = \sum_{j=1}^{Q} W_d^H \left(S_s(f_{sj}) \otimes S_t(f_{dj} + \Delta f_d) \right) \tag{5.14}$$

So far, the problem of calculating the space–time matched filter weight has been modeled as a constrained optimization problem. By solving the optimization problem above, the suppression filter, whose frequency response is matched with BiSAR clutter distribution, can be directly constructed. And then, space–time filtering can be effectively achieved in non-stationary BiSAR clutter suppression. When the objective function $F_0(W_d)$ achieves its minimum value, the optimal performance of non-stationary clutter suppression in BiSAR can be achieved.

5.2   Optimal Solution of Space–Time Filter Weight

For the COP constructed above, particle swarm optimization (PSO) algorithm [84–86] is generally applied to obtain the optimal solution of COP.

PSO algorithm is a swarm intelligence optimization algorithm, which is inspired by the biological population behavior. In the nature, birds usually randomly searching for food in the habitat. No birds know the location of food in advance, but the bird individuals can judge the distance between the its current position and food according to the smell of food. Meanwhile, through communication and information among individuals, birds can keep approaching the individual that is the closest to food. Finally, food is found by birds.

Inspired by the study on predation behavior of birds, Eberhart and Kennedy proposed the PSO algorithm in 1995 [84]. In PSO algorithm, its solution x is usually referred to as particle or individual, which can be regarded as a bird in a flock. Each particle is composed of D-dimensional decision variables, and each decision variable represents a certain independent variable in the optimization problem. In general, there is a one-to-one correspondence between the decision variables and the solution of the optimization problem in the D-dimensional space, so that the solution of the optimization problem can be continuously optimized in the process of PSO algorithm.

In PSO algorithm, the optimal solution is obtained by the cooperation and information sharing among individual particles in group. Basic and common operators of PSO include initialization (of the population), fitness function evaluation and particles property update (including the velocity and position of particles). At the beginning of each iteration, with the information sharing among particles, the global best ($gBest$) and particle best ($pBest$) can be picked by the fitness value of each particle. And then, particles are updated through tracking these two extreme values ($gBest$ and $pBest$). As iterations and updates go on, particles in the population tend to be better.

By using PSO algorithm, the COP in Eq. (5.9) could be solved. The specific steps of solving optimally space–time filter weight are given as follows:

Step 1: Identify the system parameters and configuration parameters of BiSAR, which mainly include: system carrier frequency, RPF, antenna size, flight height, flight velocity, direction, and etc.

Step 2: According to BiSAR geometrical model established in Chap. 3, for the range cell under test, the geometrical relationship between clutter scattering points and two platforms can be obtained. Then, the space–time information of clutter distribution can be acquired.

Step 3: Based on the acquired space–time information of clutter distribution, the space–time filter for BiSAR clutter suppression is designed, and the problem of calculating the filter weight is modelled as a COP.

Step 4: Apply PSO algorithm to solve the constructed COP.

4.1 Initialize parameters in PSO algorithm, including decision space V_D, dimension of particle D, number of particle Ω and the maximum iteration time G.

4.2 Initialize the particle population Γ_1. Let the current time as $g = 1$, and generate the particle population Γ_1 composed of Ω particles in decision space $V_D.x_i(g) = \left(x_i^1(g), x_i^2(g), \ldots, x_i^D(g)\right)$ denotes the $i-th$ particle in the $g-th$ generation particle population, including D independent variables. The velocity of particle in Γ_1 is set as $v_i(g) = \left(v_i^1(g), v_i^2(g), \ldots, v_i^D(g)\right)$. In this method, the initialized particle position and velocity are expressed as

$$x_i^j(1) = \min\left(V_D^j\right) + rand[0, 1] \times \left[\max\left(V_D^j\right) - \min\left(V_D^j\right)\right], j = 1, 2, \ldots, D$$

$$(5.15)$$

$$v_i^j(1) = 0, \ j = 1, 2, \ldots, D \tag{5.16}$$

where $\left\{ x_i^j(1) | j = 1, 2, \ldots, D \right\}$ is the $j - th$ independent variable contained by the $i - th$ particle in Γ_1. $\left\{ v_i^j(1) | j = 1, 2, \ldots, D \right\}$ is the $j - th$ velocity of the $i - th$ particle in Γ_1. $\min\left(V_D^j \right)$ and $\max\left(V_D^j \right)$ are the minimum and maximum value of the x_i^j in V_D. $rand[0, 1]$ is a random number between 0 and 1, which obeys uniform distribution. The initial velocity of all particles being zero.

After initialization, evaluate the fitness value of each particle, i.e., evaluate the corresponding value of the objective function $F(x_i(g))$.

4.3 When iteration time is satisfies $g \in [1, G]$, turn to step 4.4, otherwise, turn to step 5.

4.4 According to the evaluated value of the objective function, record the best value of each particle as the $pBest$ given by

$$pBest_i = F^{-1}\left[\min_g F(x_i(g)) \right] \tag{5.17}$$

where $\{ pBest_i | i = 1, 2, \ldots, \Omega \}$ is the best particle among the $i - th$ particles, and $F^{-1}[\cdot]$ is the inverse function of the objective function.

Through the information sharing among particles, the $pBest$ can be compared and the best $pBest$ can be obtained, which is taken as the current global best. The current global best is given by

$$gBest(g) = F^{-1}\left[\min_i F(x_i(g)) \right] \tag{5.18}$$

where $\{ gBest(g) | g = 1, 2, \ldots, G \}$ is the global best in $g - th$ generation particles.

4.5 Update the position and velocity of particles. First, the velocity of particles can be updated as

$$v_i(g + 1) = \kappa v_i(g) + \Delta v_i(g) \tag{5.19}$$

where $\kappa \geq 0$ is the inertia factor and $\Delta v_i(g)$ is the update variable of the $i - th$ particle in the $g - th$ generation particle population given by

$$\Delta v_i(g) = C_1 \times rand[0, 1] \times (pBest_i - x_i(g)) + C_2 \times rand[0, 1] \times (gBest(g) - x_i(g)) \tag{5.20}$$

where C_1 and C_2 are the individual learning factor and social learning factor of particles, respectively.

Based on the updated velocity information, the position of the particle can be updated:

$$x_i(g+1) = x_i(g) + v_i(g+1) \tag{5.21}$$

After the update of particle properties, the next generation of particle population Γ_{g+1} is obtained. Reevaluate the value of the fitness value of each particle in Γ_{g+1} and update iteration time as $g = g + 1$. And then, return to step 4.3.

Step 5: When the particle update process is terminated, the last generation of the particle population Γ_G can be obtained and each particle in Γ_G will gather at the location of the global optimal solution, i.e., the optimal solution of the COP in (5.9).

Step 6: According to the obtained global optimal solution, the matched space–time filter can be constructed and the strong non-stationary BiSAR clutter can be effecticely suppressed, which lays the foundation for the subsequent BiSAR-MTD processing.

In general, the variables in decision vector are real numbers. However, the filter weight W_d consists of by NK complex numbers. Thus, the filter weight will be split into real and imaginary parts. Then, the weight W_{di} can be rewritten as

$$W_{di} = x_i + jx_l, i = 1, 2, \ldots, NK \tag{5.22}$$

where the index $l = i + NK$ and the term $x = [x_1, x_2, x_3, \ldots, x_{2NK}]^T$ is one of the particles (solutions) in PSO, which consists of $2NK$ decision variables.

After implementing the proposed method, the optimal solution $x_d^* \in 2NK \times 1$ to the COP in (5.9) can be identified and then used for effective non-stationary clutter suppression in BiSAR (for detailed explanations and discussions about PSO, interested readers can refer [84]).

5.3 BiSAR Clutter Suppression via Space–Time Filtering

Through performing PSO in solving the COP given in (5.9), the matched space–time filter design can be achieved. With the obtained desired filter weight, non-stationary clutter can be effectively suppressed by matched space–time filtering in this subsection.

After range compression, BiSAR signal can be expressed as

$$S(t, f_\tau) = \omega_r(f_t)\omega_a(t) \exp\left\{-j2\pi \frac{f_\tau + f_c}{c} R_s(t, n)\right\} \tag{5.23}$$

First, to avoid the performance degradation of clutter suppression caused by RCM, the pre-processing and Keystone transform are applied to compensate the RCM as well. The pre-processing function is constructed as

$$H_{pre} = \exp\left\{-j2\pi f_{ref}\frac{f_\tau + f_c}{f_c}t\right\} \tag{5.24}$$

where the term f_{ref} is the Doppler centroid of the reference point. After multiplying H_{pre} with $S(t, f_\tau)$, RCM can be corrected by the Keystone transform given by

$$t = \frac{f_c}{f_\tau + f_c}t_1 \tag{5.25}$$

where t_1 is the new azimuth time variable after the transform. After RCM correction, the differences between clutter and target signal are mainly reflected in their azimuth phase, including Doppler frequency and spatial frequency information.

Then, to eliminate the effect of Doppler spectrum extension, the aperture time is divided into an adequate amount of time periods for finite impulse response (FIR) filtering, i.e., time division processing [87]. The vectorized signal of BiSAR signal in each period is donated by $S_{vec}(t_1)$. The length of the period ΔT is determined by the azimuth frequency modulated rate K_a and the Doppler frequency resolution δ_a. Length ΔT should satisfy the following relationship:

$$\|\Delta T \times K_a\| \le \delta_a = \frac{1}{T_{syn}} \tag{5.26}$$

where term T_{syn} denotes synthetic aperture time.

According to the obtained optimal solution x_d^*, the desired optimal filter weight $W_d^* \in NK \times 1$ can be obtained from (5.22), which is given by

$$W_{opt} = \begin{bmatrix} x_{d1}^* + jx_{d(NK+1)}^*, x_{d2}^* + jx_{d(NK+2)}^*, \\ x_{d3}^* + jx_{d(NK+3)}^*, \ldots, x_{dNK}^* + jx_{d(2NK)}^* \end{bmatrix}^T \tag{5.27}$$

For the interested range cell, the space–time filtered signal can be obtained as

$$S_{MSTF}(t) = W_{opt}^H S_{vec}(t) \tag{5.28}$$

where the operation H represents conjugate transpose operation.

By using the designed filter for BiSAR clutter suppression, the performance loss caused by the strong non-stationarity of clutter can be eliminated, since the suppression filter is constructed by directly solving the modelled COP without CCM estimation. Therefore, BiSAR non-stationary clutter can be effectively suppressed and GMT signal can be retained by the MSTF method. The flowchart of the proposed optimization-based clutter suppression method (i.e., MSTF method) is given in Fig. 5.2.

Fig. 5.2 The flowchart of the proposed optimization-based clutter suppression method

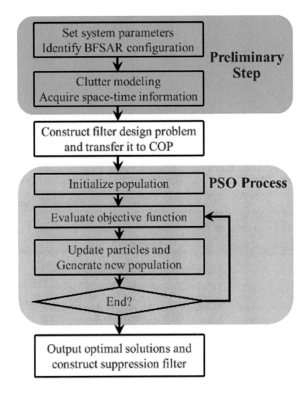

5.4 Method Validation and Performance Analysis

BiSAR clutter suppression method based on matched space–time filtering involves filter design, optimal weight solution and space–time filtering. Method validation is an important means of testing and improving method as well as evaluating its performance. The Method validation and performance analysis will be conducted from the following two aspects: example of the matched space–time filter design and example of non-stationary BiSAR clutter suppression.

5.4.1 Example of the Matched Space–Time Filter Design

In order to analyze the performance of the space–time filter obtained by solving COP in this chapter, a filter design simulation is given in this section. In the simulation, the receiver and the transmitter are considered to fly in parallel. Angles δ_T and δ_R are 90°. The transmitter has only one transmitting channel. The carrier frequency of the transmitted signal is 10 GHz, and PRF is 2000 Hz. In order to display the filter design performance of this method more intuitively, the receiver is equipped with a uniform linear array, which is placed perpendicular to the flight direction of the receiver. Both space DOF and time DOF are set to be 6. Coordinates of the receiver and the transmitter are $(0, -3000, 4000)$m and $(6000, -3000, 4000)$m, respectively. Flight velocities of platforms are 120 m/s.

With the above BiSAR configuration, an example of filter design is carried out. The obtained space–time filter by MSTF method is compared with the traditional STAP method. In the traditional STAP method, the number of secondary samples used for CCM estimation is set as $2NK + 1$, according to the RMB criterion. For the PSO algorithm, the population size X is set to be 100 and the maximum iteration times G is set to be 100. The lower and upper boundaries of the decision variables are set as -0.05 and 0.05, respectively.

Figure 5.3 shows the comparison of the filter's space–time frequency response between the traditional STAP and the proposed MSTF method. Figure 5.3a is BiSAR clutter ridge of CUT in the space–time domain. Figure 5.3b gives a space–time frequency response of the filter obtained by the traditional STAP. Due to the range

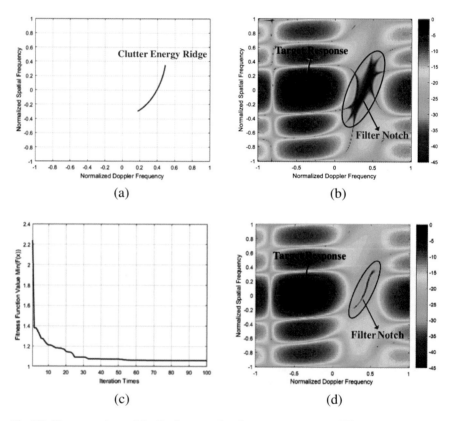

Fig. 5.3 The comparison of the filter's space–time frequency response. **a** BiSAR clutter ridge in space–time domain. **b** Filter response in traditional STAP. **c** Solving process of MSTF. **d** Filter response in the proposed MSTF technique

dependence of clutter, CCM used for filter weight calculation in traditional STAP is difficult to be correctly estimated. It can be seen from Fig. 5.3b that the filter notch is wide and does not match the BiSAR clutter ridge in space–time domain, which would lead to the performance deterioration in clutter suppression and GMT detection. The simulation results of the proposed MSTF is shown in Fig. 5.3c and d. Figure 5.3c shows the solving process, which is the relationship between the fitness function value min F (x) and iteration times. Figure 5.3d shows the 2-D response of the matched space–time filter designed by the proposed method. Compared with Fig. 5.3b, we can observe that filter notch in Fig. 5.3d is narrowed down and its space–time location is consistent with the clutter ridge of CUT. The filter in Fig. 5.3d matches clutter ridge better than that in Fig. 5.3b. Thus, clutter suppression via the proposed MSTF technique will not be impacted by the non-stationarity of clutter in BiSAR.

In order to further compare the processing performance of these two filters, the profiles of the filter response in space–time domain are shown in Fig. 5.4. The blue dashed line in the figure is the frequency response profile of the filter obtained by

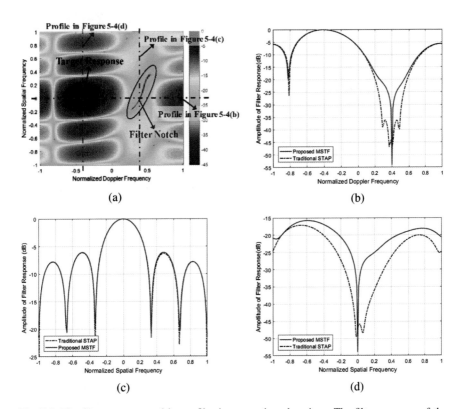

Fig. 5.4 The filter response and its profiles in space–time domain. **a** The filter response of the designed filter. **b** The profile along the Doppler frequency. **c** The profile along the spatial frequency (Normalized Doppler = −0.4). **d** The profile along the spatial frequency (Normalized Doppler = 0.4)

the traditional STAP method, and the solid red line is the frequency response profile of the filter designed by the MSTF method.

Figure 5.4a shows the way of filter profile division in space–time domain. The filter response to GMT in space–time domain is located at the position, where the normalized Doppler frequency f_d is -0.4 and spatial frequency f_s is 0. First, the filter response to GMT is analyzed. Figure 5.4b is the profile of filter response along the normalized Doppler frequency axis ($f_s = 0$) and Fig. 5.4c is the profile of filter response along the normalized spatial frequency axis ($f_d = -0.4$). From these two figures, it can be seen that the frequency response of the filter designed by the proposed MSTF method to GMT is 0 dB, which can effectively retain the energy of moving target, and provide a basis for the subsequent target detection. Second, the filter notches are compared in two dimensions. Figure 5.4 d shows the filter profile along the normalized spatial frequency axis ($f_d = 0.4$). According to the results in Fig. 5.4b and d, it can be obviously observed that the filter notch designed by the MSTF method is narrower and deeper, compared with the traditional STAP method. Thus, performing the matched space–time filtering with the filter in Fig. 5.3d, the proposed method is not affected by the strong nonstationary aspect of BiSAR clutter and it would have a better performance in BiSAR non-stationary clutter suppression than the traditional STAP method.

5.4.2 Example of Non-stationary BiSAR Clutter Suppression

In this section, an example of non-stationary BiSAR clutter suppression is given, which further validates the effectiveness of the proposed MSTF method. Figure 5.5 gives the geometrical configuration of BiSAR. The transmitter and the receiver are flying parallelly, and their flight velocities are 80 m/s. Table 5.1 lists other system parameters. To further demonstrate the effectiveness of the proposed method, clutter suppression performance of DPCA, the traditional STAP and MSTF are compared in

Fig. 5.5 The geometrical configuration of BiSAR

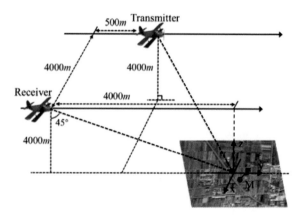

Table 5.1 System parameters

Parameters	Values
Carrier frequency	10 GHz
Bandwidth	150 MHz
PRF	1000 Hz
Space DOF	6
Time DOF	6
Transmitter location	$(-4000, -3500, 4000)$ m
Receiver location	$(0, -4000, 4000)$ m

the following. The number of antenna is 3 and 8 pulses are processed in one CPI. For the PSO algorithm, the parameters are set similar to the example of the filter design. A moving target is considered and it is denoted by MT. The motion parameters of MT is set as: $(v_x, v_y) = (-4, 4)$ m/s. Suppose MT is a car moving slowly on the country road.

Figure 5.6 shows the processing results of BiSAR signal in echo domain and image domain. Figure 5.6a is the RCM correction result in echo domain. Figure 5.6b is the imaging result of BiSAR echo. From the results in Fig. 5.6, it can be seen that MT is submerged in the clutter background and the effective GMT detection is unable to be realized, neither in echo domain nor image domain, which brings great difficulties for BiSAR-MTD.

In order to suppress BiSAR clutter and eliminate the influence on target detection, traditional DPCA method is applied first in non-stationary BiSAR clutter suppression. The suppression results in echo domain and image domain are shown in Fig. 5.7. From the results in Fig. 5.7, we can observe that DPCA method is ineffective when it comes to BiSAR. The reason is that the transmitter and receiver are located on two different platforms in BiSAR configuration, which means that the DPCA condition cannot be always satisfied and will lead to a worse suppression result. Furthermore, the equivalent phase centers of receiving channels are time-variant and spatial-variant in BiSAR, which will further increase the difficulty of clutter cancellation. Therefore, traditional DPCA is unsuitable for BiSAR clutter suppression.

Figure 5.8 shows the non-stationary BiSAR clutter suppression results via using the traditional STAP method. Compared with the suppression results in Fig. 5.7, part of non-stationary clutter could be suppressed by the traditional STAP method, and most clutter energy has been mitigated. However, due to the strong non-stationary characteristic of BiSAR clutter, the optimum performance cannot be achieved. From the results in Fig. 5.8, we can find that the residual clutter signal is still retained in BiSAR echo after STAP processing. Since BiSAR clutter can not be suppressed sufficiently, MT is hard to be distinguished from residual clutter in the following detection, which will result in plenty of false alarms and seriously deteriorate the performance of BiSAR-MTD processing.

Figure 5.9 shows the non-stationary BiSAR clutter suppression results via the proposed MSTF method. Compared with the suppression results of the above two

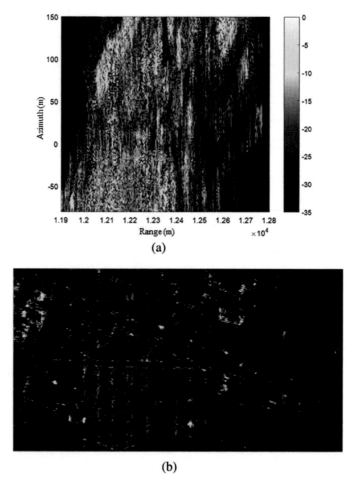

Fig. 5.6 Processing results of BiSAR echo. **a** Echo domain. **b** Image domain

traditional methods, it can be obviously seen that the MSTF method has a better
non-stationary clutter suppression performance. The proposed MSTF technique can
effectively suppress BiSAR non-stationary clutter and retain the moving target signal
as a whole. Therefore, by performing the matched space–time filtering, MT can be
easily detected and the relevant MTD processing can be effectively achieved in
BiSAR cases.

Figure 5.10 shows the azimuth and range profiles of moving target in Fig. 5.9.
In Fig. 5.10, the gray dashed line and the red solid line represent signal amplitude
before and after clutter suppression, respectively. Before space–time filtering, MT is
submerged in strong clutter and cannot be distinguished and detected directly. After
space–time filtering, BiSAR clutter has been sufficiently suppressed and MT is well
retained after suppression. The output SCNR has been improved to about 36.56 dB.

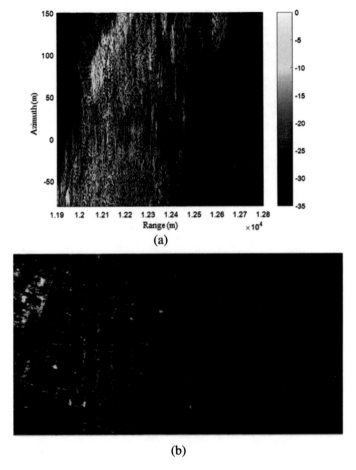

Fig. 5.7   Processing results of DPCA. **a** Echo domain. **b** Image domain

Thus, ground moving target detection in BiSAR cases can be effectively realized after clutter suppression by the proposed method, since the SCNR in detection is much higher than 13 dB [71, 88]. It can be seen from the suppression results that the proposed MSTF method in this chapter can obtain a good non-stationary clutter suppression effect, which proves the effectiveness of this method.

The comparison result among DPCA, the traditional STAP and the proposed method is shown in Fig. 5.11. It can be seen that MT is hard to be distinguished from the clutter background and it's impossible to be detected after DPCA processing. Compared with DPCA, the clutter suppression performance of STAP is better. However, since BiSAR clutter is range-dependent, residual ground clutter exists after traditional STAP processing, which leads to the performance loss in BiSAR-MTD. From the red line in Fig. 5.11, we can find that moving target MT can be easily detected after performing the proposed MSTF. The SCNR improvement of

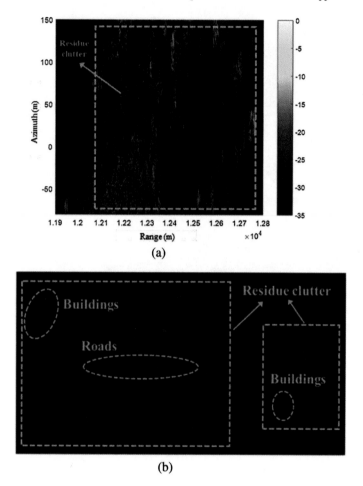

Fig. 5.8 Processing results of STAP. **a** Echo domain. **b** Image domain

the MSTF technique is about 17.02 dB more than that after STAP processing. There-fore, the proposed MSTF has a better clutter suppression performance than the most popular traditional clutter suppression methods, i.e., STAP and DPCA.

5.5 Summary

This chapter aims to address the problem of non-stationary clutter suppression with the way of solving an optimization problem. To achieve this aim, we presented an optimally matched space–time filtering technique for BiSAR. The contribution of the proposed method can be concluded as: The suppression filter is directly designed in space–time domain based on the clutter space–time characteristics, and its weight can

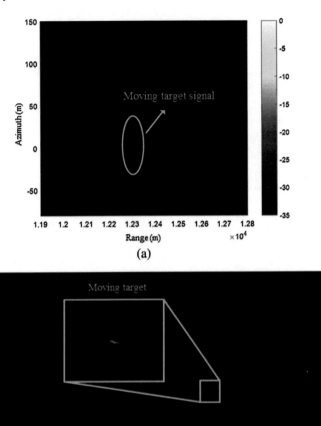

Fig. 5.9 Processing results of the proposed MSTF. **a** Echo domain. **b** Image domain

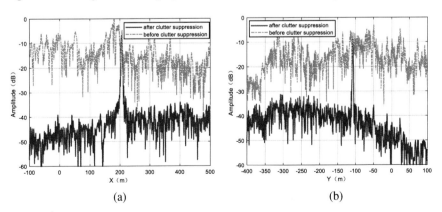

Fig. 5.10 Profiles of MT before and after clutter suppression. **a** Along x-axis. **b** Along y-axis

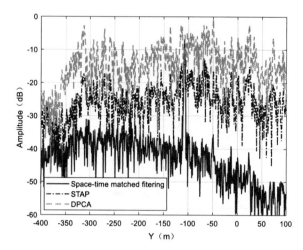

Fig. 5.11 The profile of MT after clutter suppression via DPCA, the traditional STAP and the proposed MSTF method

be obtained without CCM estimation via applying optimization algorithm. Thus, the proposed method is not affected by the strong nonstationary aspect of BiSAR clutter. Therefore, the proposed method has a better performance in BiSAR clutter suppression than the traditional STAP method. Performance analysis results are presented to illustrate the effectiveness of the proposed method about non-stationary clutter suppression. The results of the proposed method can be conveniently used to guide the space–time filter design for the effective BiSAR clutter suppression, which is highly desirable in practical applications.

Chapter 6
Spare-Recovery-Based Clutter Suppression Method

Abstract In this chapter, a clutter-ridge matched spare recovery STAP (CRM-SR-STAP) method for BiSAR non-stationary clutter suppression is proposed. First, clutter modeling with arbitrary BiSAR configuration is applied to obtain the clutter ridge accurately in space–time plane. Then, Keystone transform and time-division processing are applied to correct range cell migration and eliminate Doppler frequency migration. Next, according to the calculated clutter ridge, the CRM dictionary is reconstructed via adaptive gradient method, which is distributed along the direction of the clutter ridge and its orthogonal direction to solve the off-grid problem. Finally, the clutter covariance matrix (CCM) is estimated accurately with the preprocessed data and the CRM dictionary. And the filter is designed by utilizing the estimated CCM to suppress the non-stationary clutter. Since a few secondary samples are required in this method, the proposed method overcomes the performance degradation caused by BiSAR non-stationary inhomogeneous clutter. Simulation results are given to prove the effectiveness of this method.

Keywords Clutter suppression · Sparse recovery (SR) · Space–time adaptive processing (STAP) · Clutter-ridge matched (CRM) dictionary

6.1 The Principle of SR-STAP

For a bistatic SAR system, the receiver is equipped with N channels along the track with a spacing of d, and M pulses are transmitted during a coherent-processing-interval (CPI). Assume the contribution of a point scatter in overall BiSAR echo is $a(f_s, f_d)$ which is a complex quantity, the BiSAR clutter plus noise snapshots for a range bin can be expressed as,

$$x = \sum_{f_s, f_d} a(f_s, f_d)s(f_s, f_d) + n \tag{6.1}$$

where $s(f_s, f_d)$ denotes the space–time steering vector with spatial frequency f_s and Doppler frequency f_d, which can be given by

© The Author(s), under exclusive license to Springer Nature Singapore Pte Ltd. 2022
Z. Li et al., *Bistatic SAR Clutter Suppression*,
https://doi.org/10.1007/978-981-19-0159-1_6

$$s\left(f_{s,i}, f_{d,i}\right) = s_s\left(f_{s,i}\right) \otimes s_t\left(f_{d,i}\right)n$$
$$s_s\left(f_s\right) = \left[1, e^{j2\pi f_s}, e^{j2\pi \cdot 2f_s}, \cdots, e^{j2\pi \cdot (N-1)f_s}\right]^T \tag{6.2}$$
$$s_t\left(f_d\right) = \left[1, e^{j2\pi f_d}, e^{j2\pi \cdot 2f_d}, \cdots, e^{j2\pi \cdot (M-1)f_d}\right]^T$$

where $\otimes$ represents the Kronecker product, $[\cdot]^T$ denotes the transposition operation, $S_s(f_s)$ and $S_t(f_d)$ are the spatial steering vector ($N \times 1$) and the time steering vector ($M \times 1$). In spare recovery STAP (SR-STAP) method, the spatial-frequency plane is divided into grids by discretizing spatial and Doppler axis and $N_s = \rho_s N$, $M_d = \rho_d M$ denote their quantization grids respectively. The content in each grid represents the echo signal with spatial frequency and Doppler frequency corresponding to that grid. Then the data snapshot can be written as

$$x = \sum_{i=1}^{N_s} \sum_{j=1}^{M_d} a_{i,j} s\left(f_{s,i}, f_{d,j}\right) + n \tag{6.3}$$

The above equation can be written as matrix form as shown in Eq. (6.4),

$$x = \phi a + n, a = \left[a_{1,1}, a_{1,2}, ..., a_{N_s, M_d}\right]$$
$$\phi = \left[s\left(f_{s,i}, f_{d,1}\right), s\left(f_{s,i}, f_{d,2}\right), ..., s\left(f_{s,N_s}, f_{d,M_d}\right)\right] \tag{6.4}$$

where a denotes the amplitude of the profile of the clutter ridge at the corresponding space–time position. $\phi = \left[s\left(f_{s,i}, f_{d,1}\right), s\left(f_{s,i}, f_{d,2}\right), ..., s\left(f_{s,N_s}, f_{d,M_d}\right)\right]$ denotes the overcomplete dictionary. The above equation is the fundamental model of SR-STAP. It has three important characteristic worthy of indicating explicitly [89]. Firstly, the energy distribution of clutter and potential target buried in it in space–time domain can be obtained by solving linear equation of unknown a and known x. Secondly, the column dimension $N_s M_d$ of matrix ϕ is much greater than the row dimension NM. Hence, Eq. (6.4) is heavily ill-posed. There are infinite vectors satisfying (6.4) in general and some constraint should be imposed to get some solution with unique feature. Finally, from viewpoint of STAP, the above solution is "sparse". That is to say, most of its entries that are negligible and only a small portion of elements in solution vector, which represents contribution of main clutter and potential target, are remarkable [55, 56].

SR-STAP method requires only a few secondary samples to reconstruct the space–time spectrum, which avoids the performance degradation caused by BiSAR non-stationary clutter. In SR-STAP method, to estimate the clutter space–time profile in (6.4), Eq. (6.4) can be modeled as the following optimization problem:

$$\min_a \|a\|_0 \quad s.t. \|x - \phi_{CRM} a\|_2 \leq \varepsilon \tag{6.5}$$

where $\|\cdot\|_0$ is the l_0 norm, $\|\cdot\|_2$ is the l_2 norm, and ε is the error tolerance, which is usually set as the noise power. With the estimated space–time profile a, the clutter covariance matrix of x can be obtained by

$$\widehat{R} = E\left[xx^H\right] = \phi^H diag(P)\phi + \sigma_n^2 I_{NM}$$
$$P = [P_{1,1}, P_{1,2}, ..., P_{N_s,M_d}] \tag{6.6}$$
$$P_{i,j} = E\left[\left|a_{i,j}\right|^2\right], i = 1, 2, ..., N_s, j = 1, 2, ..., M_d$$

where σ_n^2 is the noise variance that is uncorrelated with the other components. With estimated CCM $\widehat{R}$, the weighting vector can be obtained by

$$\omega_{opt} = \mu \hat{R}^{-1} s_t \tag{6.7}$$

where μ is a normalization constant, s_t is target's space–time steering vector.

6.2 Off-Grid Problem in SR-STAP for BiSAR

Under side-looking monostatic SAR scenario, the clutter ridge can be perfectly located on the grid of the conventional dictionary, as shown in Fig. 6.1a.

However, in BiSAR scenario, the clutter ridges are no longer a straight and uniform distribution in the space-Doppler plane. The energy of the clutter will be evenly distributed on the plane grid points with a lower probability, and the energy will leak and be distributed on more grids as shown in Fig. 6.1b. It can be observed that due to the off-grid problem, compared with the monostatic SAR, the BiSAR sparsity is

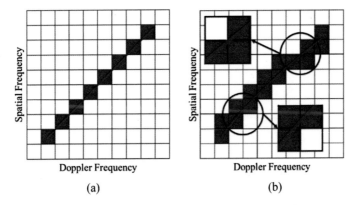

Fig. 6.1 Clutter ridge in the discretised space–time domain. **a** Monostatic SAR without off-grid problem. **b** Bistatic SAR with off-grid problem

reduced by 75%. The grid mismatch will cause the clutter energy leakage, which will broaden the clutter spectrum and affect the SR-STAP filter performance. Although the encryption grid can increase the probability of clutter falling on the grid points, it will increase the correlation between adjacent atoms in the base dictionary and reduce the recovery performance [90]. Therefore, a new dictionary construction method is needed to eliminate or mitigate the impact of grid mismatch.

6.3 CRM Dictionary Reconstruction via Adaptive Gradient Method

Based on the analysis in section II, the off-grid problem of the BiSAR clutter leads to inaccurate CCM estimated by traditional SR technique, and the filter performance is significantly degraded. To solve the off-grid problem caused by the complex space–time characteristics of BiSAR clutter, this chapter proposes a CRM dictionary reconstruction via adaptive gradient method that perfectly matches the clutter ridge. Figure 6.2 shows BiSAR clutter ridge in the space–time plane. It can be seen that its space–time distribution exhibits nonlinear characteristics. The instantaneous solution for the Doppler frequency as a function of the spatial frequency f_s may be modeled as the follows

$$f_d = F(f_s) = a_0 + a_1 f_s + a_2 f_s^2 \tag{6.8}$$

With the BiSAR clutter ridge obtained from the BiSAR space–time clutter modeling, the terms a_0, a_1, a_2 in Eq. (6.8) can be calculated via fitting the clutter

Fig. 6.2 Clutter-ridge matched dictionary reconstruction scheme

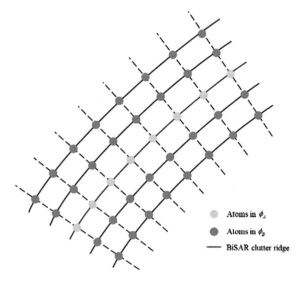

ridge. Along the direction of the clutter ridge, the atoms of the first part in CRM dictionary are constructed as follows,

$$
\phi_A = \left\{ a(f_s, f_d) \left| \begin{array}{l} a_i\left(f_{s_i}, f_{d_i}\right) = \left(f_{s_i}, F\left(f_{s_i}\right)\right) \\ f_{s_i} = f_{s_0} + len_A \cdot i - len_A \cdot n/2, i = 0, 1, \ldots, n \end{array} \right. \right\} \tag{6.9}
$$

where $F(\cdot)$ denotes the fitting function of the clutter ridge, f_{s_0} denotes the center spatial frequency of the BiSAR clutter, and len_A denotes grid length along the clutter ridge. The atoms in ϕ_A can perfectly adaptively match the clutter ridge of BiSAR in the space–time plane, ensuring the received clutter falls on the CRM dictionary grid. In addition, using the gradient of the clutter ridge, the atoms of the second part in CRM dictionary are constructed along the orthogonal direction of the BiSAR clutter ridge as follows, which are marked by green atoms in Fig. 6.2, i.e.,

$$
\phi_B = \left\{ b(f_s, f_d) \left| \begin{array}{l} b = mb_i, m \neq 0 \\ b_i\left(f_{s_i}, f_{d_i}\right) = \left(\dfrac{len_B\left(1, -F'\left(f_{s_i}\right)^{-1}\right)}{\sqrt{1 + F'\left(f_{s_i}\right)^{-2}}} \right), i = 0, 1, \ldots, n \end{array} \right. \right\} \tag{6.10}
$$

where len_B denotes the grid length along the orthogonal direction of clutter ridge.

The atoms in ϕ_B are uniformly distributed along the gradient of clutter ridge. Thus, the dictionary ϕ_{new} can be obtained with ϕ_A and ϕ_B as follows,

$$
\phi_{new} = \phi_A \cup \phi_B \tag{6.11}
$$

It is worth noting that the atoms in ϕ_{new} may exceed the space–time domain, thus it is necessary to limit the range in ϕ_{new} to ensure that the atoms in CRM dictionary do not exceed the normalized space–time domain. Besides, if the quadratic term a_2 is large, the dictionary grids along the orthogonal direction may cross each other. In this case, the space–time domain can be partially filled via applying the proposed method, and the remaining positions in spacetime domain can be filled with uniform atoms to avoid grid crossings.

$$
\begin{cases} -1 \leq f_s \leq 1 \\ -1 \leq f_d \leq 1 \end{cases} \tag{6.12}
$$

With the range limitation in Eq. (6.12), the CRM dictionary ϕ_{CRM} can be obtained. Compared with the conventional dictionary in the existing SR-STAP method, the CRM dictionary matches BiSAR clutter ridge better. The schematic diagram of the CRM dictionary is shown in Fig. 6.3, which shows that the proposed CRM dictionary effectively ensures that BiSAR clutter falls on grids in the ϕ_{CRM}. For instance, with

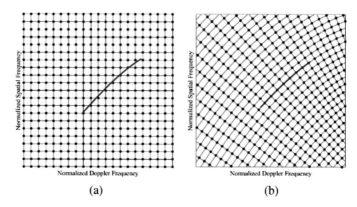

Fig. 6.3 The sketch of the existing dictionary and CRM dictionary. **a** Existing dictionary. **b** CRM dictionary

the configuration shown in Fig. 6.3b, the sketch of the existing dictionary and CRM dictionary are shown in Fig. 6.4. Compared with the existing dictionary, it can be seen that the CRM dictionary matches BiSAR clutter ridge better. Figure 6.4 shows the gradient properties of the existing dictionary and CRM dictionary, where the color scale denotes the gradient of the atoms. It can be observed that the atoms of the CRM dictionary is adaptively and uniformly distributed along the clutter ridge gradient, which matches clutter ridge perfectly.

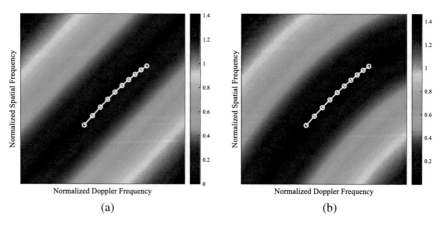

Fig. 6.4 The gradient properties of the existing dictionary and CRM dictionary. **a** Existing dictionary. **b** CRM dictionary

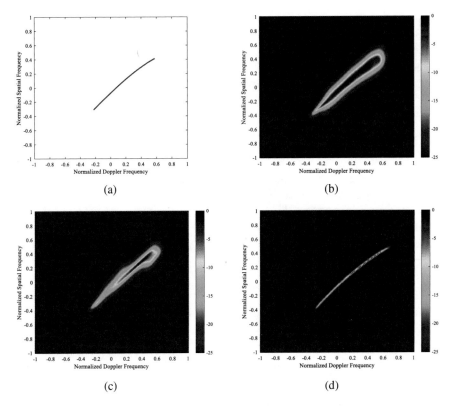

Fig. 6.5 The comparison of BiSAR clutter spectrum recovery. **a** BiSAR clutter ridge in space–time domain. **b** Estimated via traditional STAP. **c** Recovered via existing SR-STAP. **d** Recovered via CRM-STAP

6.4   The Process of the Proposed Method

Due to the range migration and non-stationarity of BiSAR clutter, it is difficult to effectively separate moving target and clutter in the space–time domain. With the non-stationarity of the BiSAR clutter, the CCM of CUT is hard to be estimated accurately from the secondary samples, which leads to the performance degradation of the traditional STAP. In order to accurately estimate CCM with a small amount of samples, the SR-STAP method has been developed. However, due to the nonlinear characteristic of clutter ridge in BiSAR cases, the off-grid problem will occur in the existing SR-STAP method, which causes the performance loss in clutter spectrum reconstruction. As a result, the non-stationary clutter cannot be suppressed effectively and moving target is difficult to be detected.

In this section, the CRM-STAP method is introduced in detail for BiSAR non-stationary clutter suppression. First, Keystone transform is applied to correct RCM. Then, clutter ridge is calculated according to the prior information, and the CRM dictionary is reconstructed via the adaptive gradient method, which is established along the clutter ridge and its orthogonal direction. In the next, with the preprocessed received data, the CCM is directly estimated by solving the MMV optimization problem with the proposed CRM dictionary. Finally, based on the estimated CCM, the CRM-STAP filter is constructed for BiSAR non-stationary clutter suppression.

6.4.1   Range Cell Migration Correction

In order to suppress BiSAR clutter effectively, the first thing is to compensate the RCM of moving target and clutter simultaneously to avoid the performance degradation in clutter suppression. To correct the RCM, the pre-processing and keystone transform are applied. The pre-processing function is given by

$$H_{pre} = \exp\left\{-j2\pi f_{ref}\frac{f_r + f_c}{f_c}t\right\}$$ (6.13)

where f_{ref} is the Doppler centroid of the reference point. After the pre-processing with the received data, the first-order Keystone transform is applied to correct the linear RCM. The transform function is

$$t = \frac{f_c}{f_\tau + f_c}t_1$$ (6.14)

where t_1 denotes the new azimuth time variable after Keystone transform. After RCM correction, the received data of the n-*th* receiving channel in range times-azimuth time domain can be expressed as

$$S_{rcmc}(t_1, \tau, n : P_c) \approx \sigma(P_c)\sin c\left[B_\tau\left(\tau - \frac{R_s(0, n; P_c)}{c}\right)\right]$$
$$\times \omega_a(t_1 - t_P)\exp\left\{-j\frac{2\pi}{\lambda}(R_s(t_1, n; P_c))\right\}$$ (6.15)

where $\sigma(\cdot)$ is the the backscattering coefficient of the scattering point P_c. τ and B_τ denotes the range time and bandwidth. The spatial frequency and Doppler frequency of clutter and moving target are included in the last exponential term in Eq. (6.15), which is the main difference between clutter and moving target.

Then in order to remove the effect of Doppler spectrum extension, the long aperture time of bistatic SAR is divided into a huge quantity of periods for finite impulse response (FIR) filtering, i.e., time division processing [91]. The vectorized signal of BiSAR echo in each period is donated by $S_{vec}(t)$. So far, the effect of RCM and Doppler spectrum extension has been eliminated for the subsequent filtering processing. However, due to the existence of the non-stationary and inhomogeneous clutter of BiSAR, moving target is generally submerged in the clutter background. Thus, the CRM-STAP method is proposed for BiSAR clutter suppression in the next subsection.

6.4.2 BiSAR Non-stationary Clutter Suppression

Generally, the existing SR-STAP method only uses a single sample, i.e. a single measurement vector (SMV) to recover the clutter spectrum, in which the information within adjacent samples are not utilized and potential performance loss may be induced. To improve the sparse recovery accuracy, MMV algorithm has been proposed, most of which are obtained by the straightforward extension of SMV algorithm, such as multiple convex optimization (MCVX) [92] and multiple orthogonal matching pursuit (MOMP) [93].

However, although MCVX and MOMP have significant improvement on CCM estimation accuracy, the computational complexity of MCVX is very high which makes the real-time implementation difficult, and MOMP may fail in the strong correlation between the atoms of the dictionary which leads to a significant reduction in recovery performance. Different from the aforementioned MMV-SR algorithm, MSBL method has a sparser solution [94], and it also has a better performance when the atomic correlation is strong. In addition, its computational complexity is more efficient than the MCVX algorithm. With the MMV model, the received data and the optimization problem can be expressed as

$$
\begin{aligned}
X &= \phi A + N \\
\min_{A} &\|A\|_{2,1} s.t. \|X - \phi_{CRM} A\| \le \varepsilon
\end{aligned}
\tag{6.16}
$$

where $X = [x^{(1)}, x^{(2)}, ..., x^{(L)}] \in C^{NM \times 1}$ denotes the received clutter signal, $A = [a^{(1)}, a^{(2)}, ..., a^{(L)}] \in R^{N_s M_d \times 1}$ denotes the clutter space–time profile matrix, $\|\cdot\|_{2,1}$ denotes $l_{2,1}$ norm, and ε is constant error tolerance.

Assuming the noise has the complex white Gaussian distribution with unknown power σ^2, we can obtain the Gaussian likelihood function of the measurement model in Eq. (6.17) as

$$
p(X|A, \sigma^2) = \frac{1}{(\pi \sigma^2)^{NML}} \exp \left[-\sigma^{-2} \sum_{l=1}^{L} \left\| x^{(l)} - \phi a^{(l)} \right\|^2 \right]
\tag{6.17}
$$

Suppose $a^{(l)}, l \in [1, L]$ is a complex Gaussian prior, which can be expressed as $a^{(l)} \sim N(0, \Gamma)$. $0 \in C^{NM \times 1}$ is a zero vector, Γ is the diagonal matrix constructed by the hyperparameter $\gamma = [\gamma_1, ..., \gamma_{N_s M_d}]^T \in C^{N_s M_d \times 1}$, which denotes the unknown variance parameters corresponding to all atoms in the overcomplete dictionary. Thus, with combining each column, the prior of A can be given by

$$p(A|\Gamma) = \pi^{-N_s M_d L} \Gamma^{-L} \exp\left(-\sum_{l=1}^{L} a^{(l)H} \Gamma^{-1} a^{(l)}\right) \tag{6.18}$$

By combining likelihood Eq. (6.17) with the prior Eq. (6.18), the posterior density of A can be obtained as

$$p(A|X, \Gamma, \sigma^2) = \frac{p(X|A, \sigma^2)p(A|\Gamma)}{\int p(X|A, \sigma^2)p(A\Gamma)dA} \tag{6.19}$$

which is modulated by the hyperparameter vector γ and σ^2. Then, the sparsity profile estimation of A is transformed into the estimation of hyperparameter γ. To estimate the hyperparameter γ, the evidence maximization (EM) algorithm is applied [95]. First, treating the unknown matrix A is treated as a hidden variable, and calculating the posterior density with known hyperparameters γ and σ^2 at the E-step. Then, γ and σ^2 are updated with the known posterior density of A at the M-step. Finally, E-step and M-step are repeated until the convergence condition is satisfied. With the recovered clutter space–time profile matrix, the CCM can be estimated by

$$\hat{R} = E[XX^H] = \phi_{CRM}^H \text{diag}(P)\phi_{CRM} + \sigma_n^2 I_{NM}$$
$$P = \|A^*\|_{2,1}, P \in R^{N_s M_d \times 1} \tag{6.20}$$

where σ_n^2 is the noise variance that is uncorrelated with other components. The weighting vector of the CRM-STAP processor can be obtained as

$$\omega_{opt} = \mu \hat{R}^{-1} s_t \tag{6.21}$$

where μ is the normalization constant, s_t is target's space–time steering vector. For the interested range cell, the CRM-STAP filtered signal can be obtained as

$$S_{CRM}(t) = (\omega_{opt})^H S_{vec}(t) \tag{6.22}$$

The proposed dictionary is reconstructed strictly according to the BiSAR clutter ridge, which effectively avoids the performance degradation caused by the off-grid problem. Therefore, CCM can be effectively estimated with a small number of samples through the MSBL algorithm, and the non-stationary inhomogeneous clutter can be effectively suppressed in BiSAR through the proposed CRM-STAP method. The procedure of the proposed method for BiSAR non-stationary clutter suppression is given in Algorithm.

The steps of the proposed method is given:

Step 1: Set the radar system parameters. Establish BiSAR clutter model of the interested range cell, and obtain the clutter ridge.

Step 2: With the obtained clutter ridge, reconstruct the clutter-ridge matched dictionary ϕ_{CRM} by the proposed method.

Step 3: Initialize the hyperparameters $\gamma := 1, \sigma_n^2 = 0.1$

Step 4: Calculate the mean μ_{k+1} and variance D_{k+1} of the unknown matrix A during the *k-th* iteration as follows

$$\mu_{k+1} = \left[\mu_{k+1}^{(1)}, \mu_{k+1}^{(2)}, \cdots, \mu_{k+1}^{(L)} \right] = \Gamma \phi^H \Sigma^{-1} X$$
$$D_{k+1} = \Gamma - \Gamma \phi^H \Sigma^{-1} \phi \Gamma$$

where $\Gamma = \text{diag}(\gamma_k), \Sigma = \sigma_l^2 I + \phi \Gamma \phi^H$.

Step 5: Update the hyperparameters γ and noise variance σ_k^2 using EM rule during the *k-th* iteration

$$\gamma_{n,k+1} = \frac{1}{L} \left[\sum_{l=1}^{L} \mu_{n,k+1}^{(l)} \right]^2 + D_{n,n,k+1}$$

$$\sigma_{k+1}^2 = \frac{\|X - \phi \mu_{k+1}\|_F^2}{L \left[NM - N_s M_d + \sum_{i=1}^{N_g M_d} \left(D_{i,i}/\gamma_i \right) \right]}$$

Step 6: Continue step 6 and step 7 until the convergence condition $\|\gamma_{k+1} - \gamma_k\|_2 / \|\gamma_k\|_2 \leq \delta$ or $\sigma^2 \leq (\sigma_*)^2$ where δ and σ_*^2 are small enough (positive thresholds) is met or the maximum number of iterations is reached.

Step 7: Assume $A^* \simeq \mu^*$, where $\mu^* \triangleq E[A|X, \Gamma^*, \sigma_*^2]$.

Step 8: Estimate the CCM $\hat{R} = \phi_{CRM}^H \text{diag}(P) \phi_{CRM} + \sigma_n^2 I_{NM}$.

Step 9: Calculate the CRM-STAP filter weight $\omega_{opt} = \mu \hat{R}^{-1} s_t$

6.5 Performance Analysis

In this section, simulations results are carried out to verify the effectiveness of the proposed CRM-STAP method for BiSAR non-stationary clutter suppression. The detailed simulation parameters are listed in Table 6.1.

Table 6.1 Simulation parameters

Parameter	Symbols name	Values
Carrier frequency	f_c	10 Ghz
Range bandwidth	B_τ	150 MHz
Pulse repetition frequency	f_r	1000 Hz
Antenna number	N	3
Pulse number	M	8
Receiver position	(x_R, y_R, z_R)	(0,–4000,4000) m
Transmitter position	(x_T, y_T, z_T)	(–4000,–3500,4000) m
Receivers velocity	V_R	100 m/s
Transmitters velocity	V_T	100 m/s

6.5.1 Clutter Spectrum Recovery Analysis

In this subsection, with system parameters in Table 6.1, the comparisons among the traditional STAP, the existing SR-STAP and the proposed CRM-STAP are given. For the existing SR-STAP method, the discrete atoms in overcomplete dictionary are uniformly sampled in both spatial and Doppler axis with a width of 0.1. The atoms of the CRM dictionary can be obtained, and the grid length len_A and len_B are set to be 0.1 as well.

Figure 6.5b shows the non-stationary clutter spectrum, which corresponds to the CCM estimated by 2MN samples in traditional STAP. It can be seen that due to the range dependence of clutter, BiSAR clutter spectrum is extended, which will cause performance degradation in clutter suppression. The recovered clutter space–time spectrum of the existing SR-STAP and CRM-STAP are shown in Fig. 6.5c and d. The comparison shows that clutter spectrum of the existing SR-STAP is broadened as well with the off-grid problem, which will degrade the performance of the SR-STAP. Since the CRM dictionary well matches clutter ridge, clutter spectrum can be recovered more accurately after CRM-STAP. To further analyze the performance of CRM-STAP, the SCNR loss is used for comparison, which is defined as,

$$SCNR_{loss} = \frac{\sigma_n^2 |\omega^H s|^2}{MN |\omega^H R\omega|} \qquad (6.23)$$

The evaluation of the SCNR loss is shown in Fig. 6.6. The results indicate that the SCNR loss of the CRM-STAP filter is closer to the optimal filter. Due to the non-stationarity of BiSAR clutter, the performance of the traditional STAP filter is seriously affected, and the performance of the existing SR-STAP filter is degraded because of the off-grid-problem. Specifically, as shown in Fig. 6.6, when the normalized Doppler frequency is 0.4, the SCNR loss improvement of the CRM-STAP is

Fig. 6.6 SCNR loss performance comparisons of proposed method and optimal case

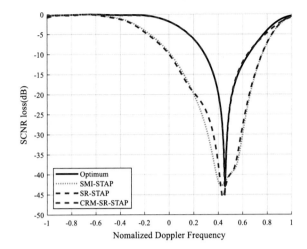

about 10 dB and 15 dB more than the existing SR-STAP and traditional STAP respectively.

6.5.2 Simulation Results of BiSAR Clutter Suppression

To demonstrate the effectiveness of this method, simulated clutter suppression results via traditional STAP, the existing SR-STAP and the proposed CRM-STAP are presented in the following. The geometric configuration is shown in Fig. 6.7 and the detailed simulation parameters are listed in Table 6.1. The number of antennas is 3, 8 pulses are processed in one CPI, and the grid widths of the existing SR-STAP and CRM-STAP methods are both set as 0.1. A moving target is considered and it is denoted by M. The motion parameter of M is set as: $(v_x, v_y) = (-4, 4)$ m/s. Suppose that M is a car whose radar cross section (RCS) is set to be 11 dB [88]. The RCS of ground, river and buildings is generally set to be 5 dB, -20 dB and 20 dB [88, 96].

Fig. 6.7 Geometrical configuration in the BiSAR simulation

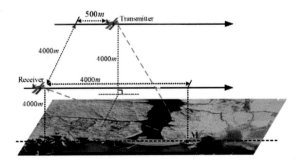

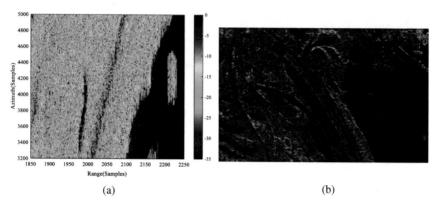

Fig. 6.8 Processing results without clutter suppression. **a** BiSAR data after range cell migration correction. **b** Imaging result

Figure 6.8 shows the received BiSAR data after range cell migration correction in range time-azimuth time domain and the imaging result without clutter suppression. It can be seen that M is submerged in the clutter background, and it is hard to be effectively detected. Thus, effective clutter suppression is necessary for the moving target detection.

The clutter suppression results in echo domain and image domain after traditional STAP, existing SR-STAP and CRM-STAP are shown in Fig. 6.9, respectively. The color scale in Fig. 6.9 is in dB. Figure 6.9a and b respectively show the results in echo domain and image domain after the traditional STAP processing. It can be seen that due to the non-stationarity of clutter, a large amount of clutter cannot be effectively suppressed by traditional STAP. Figure 6.9c and d respectively show the results in echo domain and image domain after existing SR-STAP. Taking the advantage of the CCM estimation with a few number of secondary samples, SR-STAP has a better performance than STAP. However, due to the inaccurate CCM estimation caused by the off-grid problem, there are still residual clutter and moving target is hard to be detected as well. Clutter suppression results in echo domain and image domain after CRM-STAP are shown in Fig. 6.9e and f. Compared with the above results, the non-stationary clutter can be effectively suppressed by the proposed method, and moving target can be easily distinguished. Therefore, moving target detection can be effectively realized via CRM-STAP.

To further demonstrate the clutter suppression performance of the proposed CRM-STAP method, SCNR before and after clutter suppression are analyzed. In Fig. 6.10, the gray dashed line and the red solid line respectively represent the profiles of M before and after clutter suppression. It can be seen that M is effectively retained after the proposed method, and SCNR is increased to 32.28 dB. Since it is much higher than 13 dB, M can be effectively detected [71]. Figure 6.11 shows the clutter suppression results after STAP, SR-STAP and CRM-STAP. It can be seen that SCNR of M is still less than 13 dB after STAP and SR-STAP, because of the non-stationarity of the clutter. In addition, there are a lot of residual clutter, so moving target cannot be

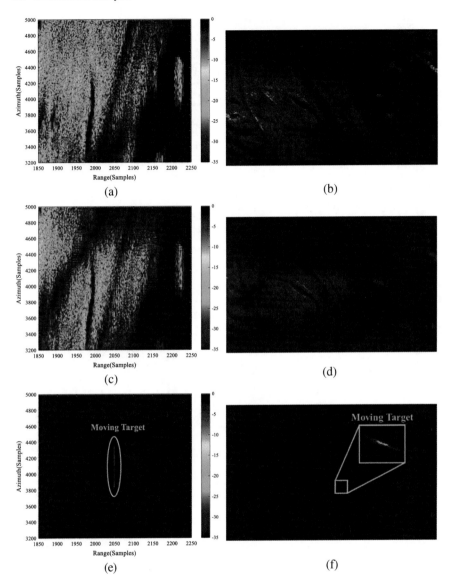

Fig. 6.9 Clutter suppression results. **a** Traditional STAP results in echo domain. **b** Traditional STAP results in image domain. **c** Existing SR-STAP results in echo domain. **d** Existing SR-STAP results in image domain. **e** Proposed CRM-STAP results in echo domain. **f** Proposed CRM-STAP results in image domain

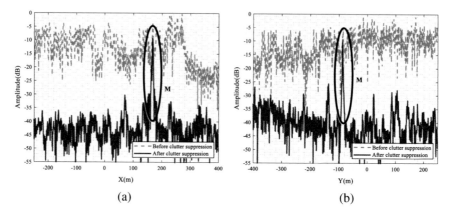

Fig. 6.10 The profiles of M before clutter suppression after clutter suppression. **a** Profile of M along the X-axis. **b** Profile of M along the Y-axis

Fig. 6.11 The profile of M after clutter suppression via the traditional STAP, the existing SR-STAP and the proposed CRM-STAP method

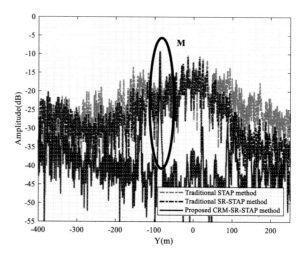

effectively detected after clutter suppression. Compared with STAP, SR-STAP has better performance. However, due to the off-grid problem, the CCM estimation is still inaccurate, resulting in clutter suppression performance degradation. The red solid line in Fig. 6.11 represents the profile of moving target after CRM-STAP. Compared with STAP and SR-STAP, SCNR after CRM-STAP processing is improves about 15 dB and 12 dB, respectively. Obviously, these results demonstrate the effectiveness of the proposed method in non-stationary inhomogeneous clutter suppression for BiSAR.

6.6 Summary

SR-STAP algorithm utilize the sparsity of clutter in space–time domain to obtain high-quality clutter space–time spectrum with a small number of samples. However, due the complex characteristic of BiSAR clutter in space–time domain, the existing sparse recovery dictionary will be mismatched with the clutter ridge, which is called off-grid problem. Because of the off-grid problem, the CCM estimated by existing SR-STAP is inaccurate, resulting in the degraded performance of clutter suppression.

In this chapter, a clutter-ridge matched STAP method for BiSAR non-stationary clutter suppression is proposed. Based on the clutter ridge calculated by the prior information, atoms in CRM dictionary are reconstructed via the adaptive gradient method, which is established along the clutter ridge direction and its orthogonal direction. Therefore, the off-grid problem can be well addressed. Finally, based on the MSBL algorithm, MMV optimization is solved and a CRM-STAP filter is constructed to suppress the BiSAR non-stationary clutter. Contributions of the proposed method include the following two aspects: (1) The constructed CRM dictionary matches the clutter ridge of BiSAR, so it can avoid the performance degradation in clutter suppression caused by the off-grid problem. (2) As the proposed method requires only a few secondary samples, it is not affected by the strong non-stationary characteristics of BiSAR clutter in heterogeneous environments. Simulation and experimental results are carried out to show the validation of the proposed method, where BiSAR non-stationary clutter has been effectively suppressed.

Chapter 7
Experimental Technique

Abstract This chapter introduces the experimental technique of BiSAR-MTD of University of Electronic Science and Technology of China (UESTC). In this chapter, BiSAR system architecture, experimental scheme, experimental results of BiSAR imaging and clutter suppression are detailed. In October 2020, the first airborne BiSAR-MTD experiment in the world was successfully conducted by UESTC. The situation of this experiment is introduced in this chapter as well and the experimental data is applied to validate the effectiveness of the method proposed in Chaps. 5 and 6.

Keywords BiSAR experiment · MTD experiment · System architecture · Experimental scheme · Experimental results

7.1 Experimental System

The whole bistatic radar system consists of two radar subsystems: transmitting subsystem and receiving subsystem. The composition of the transmitting subsystem is basically the same as that of the receiving subsystem. The difference is that the transmitting subsystem has only one antenna, which can transmit and receive radar signals, while the receiving subsystem has three antennas, which can only receive radar signals. The receiving subsystem has three mixer, three analog-to-digital converter and three memory cards. The receiving subsystem reserves the output interface of radar analog source for transmitting IF signal, and can be equipped with a mixer, fixed amplifier and antenna for transmitting radar signals.

The radar system composes of a radar analog source, an RF link, a SAR antenna, a servo drive system, a high precision synchronous source, GPS and an inertial navigation. The whole system is shown in Fig. 7.1.

Radar analog source mainly performs the functions of generation of radar signal source, acquisition of radar echo, storage of echo data, and generation of the control signal.

The RF link is responsible for up or down frequency conversion, amplification, and filtering of the linear FM signal that needs to be sent or received, so as to provide the linear FM signal or the echo signal that meets the needs for transmitting or

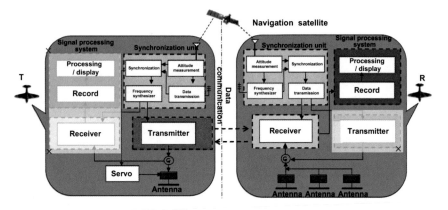

Fig. 7.1 The diagram of the whole system

receiving the link. The module is also responsible for the internal calibration of the radar signal.

GPS and inertial navigation equipment measure and determine equipment positions, attitudes, and other information. These data will be sent back to the radar central control system, and by the radar analog source equipment, GPS information will then be transmitted into the radar echo data.

The high-precision synchronization source provides the synchronization clock signal locked by GPS for transmitting subsystem and receiving subsystem to ensure that these two subsystems have certain synchronization.

The SAR antenna is located outside the onboard equipment and is used to transmit and receive radar signals.

The antenna pointing device completes the fine-tuning of the transmitting angle of the transmitting antenna to ensure that the transmitting antenna and the receiving antenna point to the same area.

The onboard equipment of the flight platform uses 220 V AC power inverted by the battery pack provided by the project to ensure a 2-h power supply.

7.2 Experimental Scheme

7.2.1 Flight Path

Considering the climate, geography and other conditions of Yinchuan, Ningxia, and aiming at stable bistatic flight and stable acquisition of scene echo data, the bistatic experiment flight path is set as follows, and the specific flight path includes the following steps:

(1)   The transimitting station follows the red line as the track in the inner ring, flying clockwise at a speed of 260 km/h.

(2)   The receiving station flies clockwise along the yellow line in the outer ring at a speed of 260 km/h.

(3)   The receiving station will take off from Yinchuan General Aviation Airport prior to the transimitting station. The receiving station will climb to the flight altitude of 2100 m and will enter the route from red 1A—the starting point of the first line.

(4)   The transimitting station will climb to the flight altitude of 1600 m, and enter the route from the starting point of the first line—yellow 1A, and both receiver and transmitter planes entered the first line from their respective 1A points at the same time.

(5)   Both receiver and transmitter aircrafts keep smooth flights within the first-line track, and the relative position relationship remains unchanged, while pulling out of the first-line from the terminus 1B.

(6)   The receiver and the transmitter enter the second line in the same way, and keep smooth flights in the second line until exiting the second line, flying total 4 laps.

(7)   1A-1B and 2A-2B as shown in the figure are straight-line flight segments and valid segments representing data admission.

(8)   Flight formation, speed and other adjustments are carried out within the dashed line segment.

When the receiver and the transmitter fly to the specified altitude, the crew on board turn on the receiver and the transmitter systems and collect echo data. Bistatic SAR-MTD experiment flight line corresponds to the main field of the scenic areas, which are (1) tarmac and runway in Yinchuan airport, and (2) farmlands and villages in Helan Country (Fig. 7.2).

7.2.2 Movement Routes of Moving Targets

In the geographical view of Yinchuan, Ningxia, moving target movement routes are set in combination with the flight path of the aircraft to ensure that targets can be irradiated by the bistatic SAR beam at the same time. The movement routes of the set moving targets are shown as follows, and the specific movement plan includes following steps (Fig. 7.3):

(1)   Target1 moves from point A to point B on the patrol lane in Yinchuan General Aviation Airport.

(2)   Target2 moves from point C to point D on the highway outside Yinchuan General Aviation Airport.

(3)   When the two platforms enter the starting point 1A of the first line, target1 and target2 begin to move in accordance with the planned route until the two platforms drive away from the first line.

Fig. 7.2   Flight path in the experiment

Fig. 7.3   Planned routes of moving targets in the experiment

(4)   If the moving target reaches the end of their planned route, it will turn around
      and continue driving until the two platforms leave the route.
(5)   After the two platforms drive away from the line, the moving targets quickly
      return to the starting point of their planned routes, that is, target1 return to point
      A, and target2 return to point C.
(6)   After moving targets back to the starting point of their planned routes, they
      will start to move when the next two stations re-enter the starting point 1A of
      the line.

7.2.3 Experiment Situation

In October 2020, we carried out the airborne bistatic SAR-MTD experiment in
Yinchuan, China. To authors' knowledge, this is the first bistatic SAR-MTD experiment in the world that both the transmitter and the receiver are mounted on aircraft
platforms.

In the bistatic SAR-MTD experiment, the airborne bistatic SAR system works in
the X-band. The airplanes in the experiment are shown in Fig. 7.4, both the receiver
platform and the transmitter platform are Cessna 208 planes. The receiver works with
forward-looking mode and the transmitter works with squint mode. The bandwidth
of the transmitted signal is 300 MHz and the PRF is 3000 Hz. Both the velocities of
the platforms are about (0, 70, 0) m/s.

In the experiment, multiple moving targets with different RCS and motion state
are considered. Two cars are selected as the targets moving in the observation area.
The selected cars are a pickup truck and a battery car as shown in Fig. 7.5. Since
only the frame of the battery car is made of metal, the RCS of the battery car is much
lower than that of the pickup target. In the experiment, the pickup truck is driven at
a speed of 20 km/h, and the battery car is driven at a speed of 25 km/h.

The following figures show the actual photos taken during the flight of the bistatic
SAR-MTD experiment. Figure 7.6a shows the flight situation of the receiver and
the transmitter taken by the ground experiment personnel, and Fig. 7.6b shows the
transmitter taken by the experiment personnel at the receiver.

The following Fig. 7.7a and b are the optical image and the bistatic SAR imaging
result of Yinchuan farmland, respectively. It can be seen from the figures that the

Fig. 7.4 The platforms in the airborne bistatic SAR-MTD experiment

Fig. 7.5 Moving targets in the airborne bistatic SAR-MTD experiment. **a** Pickup truck target; **b** battery car target

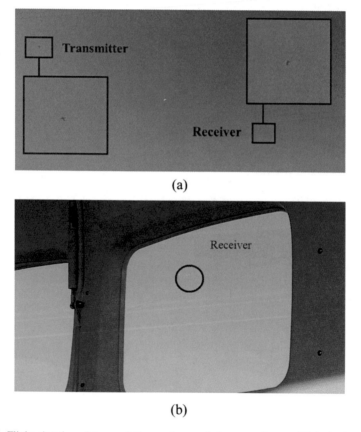

Fig. 7.6 Flight situation pictures of the receiver and the transmitter. **a** Flights' photo in the experiment; **b** The receiver observed from the transmitter

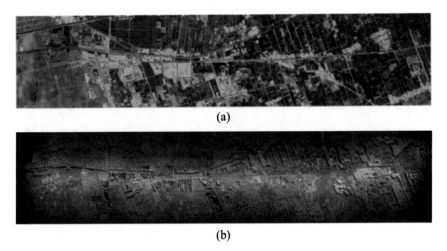

(a)

(b)

Fig. 7.7 Optical image and bistatic SAR image. **a** Optical image; **b** Bistatic SAR image

bistatic SAR image of this experiment has good imaging quality. From the picture, we can clearly see objects such as houses, roads, and trees.

7.3 Processing Results with Proposed Methods

7.3.1 Processing by Optimization-Based Method

Figure 7.8 shows the clutter suppression result processed by the proposed optimization-based method in Chap. 5 [97]. The color scale is in dB. From a and d, it can be seen that, both the pickup truck target signal and the battery car target signal are submerged in BiSAR clutter after RCMC. The signals received from the strong clutter scattering in Fig. 7.8a and d will lead to a worse detection performance in echo domain and may result in false-alarm. Figure 7.8b and e are the suppression results via MSTF for the two targets. After the matched space–time filtering with the designed filter, BiSAR nonstationary clutter has been well suppressed and the moving target signals are retained in echo domain, which can be directly used in the following target refocusing and relocation for BiSAR-MTD. To further verify the validity of the proposed method, the profiles of moving targets after clutter suppression are shown in Fig. 7.8c and f. According to the results, the output SCNR of the pickup truck target is about 15.78 dB, and that of the battery car target is about 13.96 dB, which are high enough for the following target detection in BiSAR. Therefore, the proposed method is effective for BiSAR-MTD when multiple targets exist with different motion states and RCS in the observation area, which is validated by the experimental results.

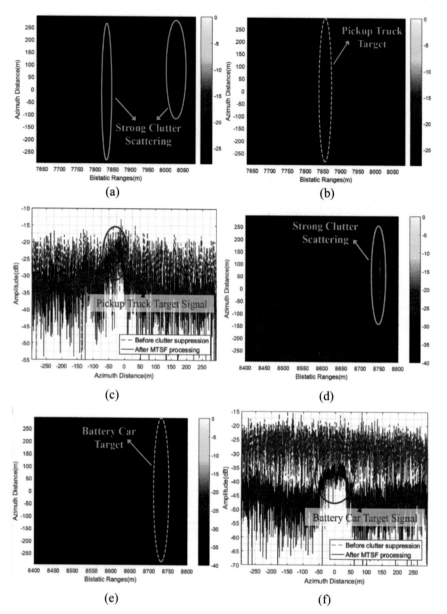

Fig. 7.8 The BiSAR-MTD experimental results with the proposed optimization-based method in Chap. 5. **a** Pickup truck signal before clutter suppression. **b** Pickup truck signal after clutter suppression. **c** The profile of the pickup truck signal. **d** Battery car signal before clutter suppression. **e** Battery car signal after clutter suppression. **f** The profile of the battery car signal

7.3.2 Processing by SR-Based Method

Figure 7.9 shows the clutter suppression result processed by the proposed CRM-based method in Chap. 6 [98]. After RCM correction, the echo data in range time-azimuth time are illustrated in Fig. 7.9a, where the color scale is in dB. In Fig. 7.9a, as the power of the clutter is greater than that of the pickup truck, the pickup truck target is submerged in the clutter and cannot be detected effectively. According to the proposed method, the received data after clutter suppression is revealed in Fig. 7.9b. It can be seen that BiSAR non-stationary clutter has been well suppressed and the moving target is still retained in echo domain. To further verify the effectiveness of the proposed method, the profiles of the pickup truck before and after clutter suppression are shown in Fig. 7.9c, where the gray dashed line and the red solid line respectively represent the profiles of M before and after clutter suppression. According to the

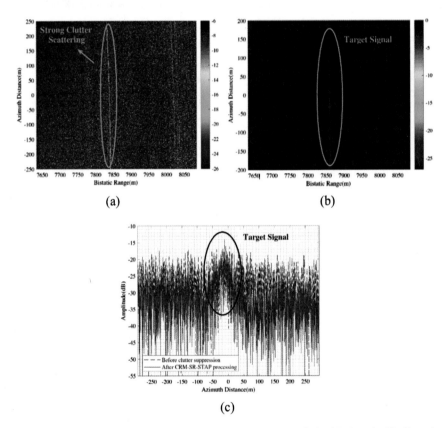

Fig. 7.9 The BiSAR-MTD experimental results with the proposed CRM-STAP method in Chap. 6. **a** Pickup truck signal before clutter suppression. **b** Pickup truck signal after clutter suppression. **c** The profile of the pickup truck signal

Fig. 7.10 Image result of pickup signal with the proposed CRM-STAP method in Chap. 6

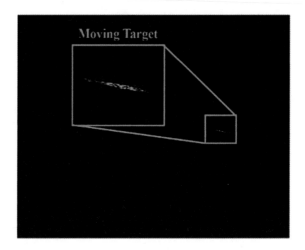

results, the output SCNR of the pickup truck target is about 15.46 dB, which can be effectively detected.

Meanwhile, Fig. 7.10 shows the processing result after clutter suppression with the proposed CRM-STAP method in image domain. It can be seen that the non-stationary clutter has been effectively suppressed by the proposed method, and the moving target can be well distinguished. Thus, the proposed method is effective for BiSAR non-stationary clutter suppression.

7.4　Summary

In order to verify the effectiveness of the proposed bistatic SAR clutter suppression method, UESTC carried out and completed the world's first airborne bistatic SAR-MTD experiment in October 2020. Firstly, this chapter introduces the composition of bistatic SAR-MTD system. Bistatic SAR is composed of a transmitting radar system and a receiving radar system, including a signal processor, a synchronization system, attitude measurement, a positioning system, and etc. Then, the scheme of the experiment is introduced, including the flight path of the carrier, the movement routes of moving targets and the experiment situation. The high-quality imaging result of bistatic SAR on Yinchuan farmland is given, which shows the effectiveness and reliability of the experiment system. Finally, the clutter suppression method based on optimization and sparse recovery is used to process the experimental data of bistatic SAR-MTD. The processing results show that both the two methods achieve the effective bistatic SAR non-stationary clutter suppression, and moving target signal is well retained. The effectiveness of the methods proposed in Chaps. 5 and 6 has been verified by the experimental data processing.

References

1. Z. Li, J. Wu, Q. Yi, Y. Huang, J. Yang, Y. Bao, Bistatic forward-looking SAR ground moving target detection and imaging. IEEE Trans. Aerosp. Electron. Syst. **51**(2), 1000–1016 (2015)
2. Z. Li, J. Wu, Y. Huang, Z. Sun and J. Yang, Ground-moving target imaging and velocity estimation based on mismatched compression for bistatic forward-looking SAR. IEEE Trans. Geosci. Remote Sens. **54**(6), 3277–3291 (2016)
3. J. Yang, *Bistatic Synthetic Aperture Radar* (National Defense Industry Press, Beijing, 2017)
4. Z. Yang, Y. Zhang, Y. Luo, *Bistatic(Multistatic) Radar Systems* (National Defense Industry Press, Beijing, 1998)
5. M. Cherniakov, *Bistatic Radar: Emerging Technology* (John Wiley and Sons Ltd., Chichester, 2008)
6. J. Balke, SAR image formation for forward-looking radar receivers in bistatic geometry by airborne illumination, in *IEEE Radar Conference, Rome* (2008), pp. 1–5
7. J. Ender, J. Klare, I. Walterscheid, et al., Bistatic exploration using spaceborne and airborne SAR sensors: A close collaboration between FGAN, ZESS and FOMAAS, in *IEEE International Symposium on Geoscience and Remote Sensing, Denver* (2006), pp. 1828–1831
8. J. Klare, I. Walterscheid, A. Brenner, et al., Evaluation and optimisation of configurations of a hybrid bistatic SAR experiment between Terra SAR-X and PAMIR, in *IEEE International Symposium on Geoscience and Remote Sensing, Denver* (2006), pp. 1208–1211
9. T. Fritz, C. Rossi, N. Yague-Martinez, et al., Interferometric processing of Tan DEM-X data, in *IEEE International Geoscience and Remote Sensing Symposium, Vancouver* (2011), pp. 2428–2431
10. J.H.G. Ender, Signal theoretical aspects of bistatic SAR, in *IEEE International Geoscience and Remote Sensing Symposium, Toulouse* (2003), pp. 1438–1441
11. M. Murray, Taking reconnaissance to another level (Sandia National Laboratories, 2004)
12. J. Balke, Field test of bistatic forward-looking synthetic aperture radar, in *IEEE International Radar Conference, Arlington* (2005), pp. 424–429
13. I. Walterscheid, T. Espeter, J. Klare, et al., Potential and limitations of forward-looking bistatic SAR, in *IEEE International Geoscience and Remote Sensing Symposium, Honolulu* (2010), pp. 216–219
14. I. Walterscheid, T. Espeter, J. Klare, et al., Bistatic spaceborne-airborne forward-looking SAR, in *European Conference on Synthetic Aperture Radar, Aachen* (2010), pp. 1–4
15. T. Espeter, I. Walterscheid, J. Klare et al., Bistatic forward-looking SAR: Results of a spaceborne–airborne experiment. IEEE Geosci. Remote Sens. Lett. **8**(4), 765–768 (2011)
16. I. Walterscheid, A.R. Brenner, J. Klare, Radar imaging with very low grazing angles in a bistatic forward-looking configuration, in *IEEE International Geoscience and Remote Sensing Symposium, Munich* (2012), pp. 327–330

17. I. Walterscheid, B. Papke, Bistatic forward-looking SAR imaging of a runway using a compact receiver on board an ultralight aircraft, in *International Radar Symposium, Dresden*, vol. 1 (2013), pp. 461–466
18. Y. Huang, J. Yang, J. Wu et al., Precise time frequency synchronization technology for bistatic radar. J. Syst. Eng. Electron. **19**(5), 929–933 (2008)
19. L. Xian, J. Xiong, Y. Huang, et al., Research on airborne bistatic SAR squint imaging mode algorithm and experiment data processing, in *Asian and Pacific Conference on Synthetic Aperture Radar, Huangshan* (2007), pp. 618–621
20. J. Chen, J. Xiong, Y. Huang, et al., Research on a novel fast backprojection algorithm for stripmap bistatic SAR imaging, in *Asian and Pacific Conference on Synthetic Aperture Radar, Huangshan* (2007), pp. 622–625
21. J. Yang, Y. Huang, H. Yang, et al., A first experiment of airborne bistatic forward-looking SAR-Preliminary results, in *IEEE International Geoscience and Remote Sensing Symposium, Melbourne* (2013), pp. 4202–4204
22. R.K. Raney, Synthetic aperture imaging radar and moving targets. IEEE Trans. Aerosp. Electron. Syst. **3**, 499–505 (1971)
23. A. Freeman, Simple MTI using synthetic aperture radar, in *IEEE International Geoscience and Remote Sensing Symposium, Strasbourg* (1984), pp. 65–70
24. H.C. Chen, C.D. Mcgillem, Target motion compensation by spectrum shifting in synthetic aperture radar. IEEE Trans. Aerosp. Electron. Syst. **28**(3), 895–901 (2002)
25. J.R. Fienup, Detecting moving targets in SAR imagery by focusing. IEEE Trans. Aerosp. Electron. Syst. **37**(3), 794–809 (2001)
26. J.R. Moreira, W. Keydel, A new MTI-SAR approach using the reflectivity displacement method. IEEE Trans. Geosci. Remote Sens. **33**(5), 1238–1244 (2002)
27. F. Zhou, Z. Li, Z. Bao, A new approach to ground moving target detection and location based on two-look processing for the single channel SAR system. J. Xidian Univ. **33**(5), 673–677 (2006)
28. S. Liu, Y. Yuan, S. Mao, The method of sub-apertures space-frequency processing for the single-channel SAR moving target detection. J. Electron. Inf. Technol. **32**(8), 1992–1996 (2010)
29. S. Barbarossa, A. Farina, Detection and imaging of moving objects with synthetic aperture radar. Part 2: Joint time-frequency analysis by Wigner-Ville distribution. IEE Proc F Radar Signal Process. **139**(1), 79–88 (1992)
30. S. Barbarossa, A. Farina, A novel procedure for detecting and focusing moving objects with SAR based on the Wigner-Ville distribution, in *IEEE International Radar Conference, Arlington* (1990), pp. 44–50
31. Y. Mao, G. Chen, A SAR/ISAR multi moving target detection method based on WVD-HT. J. Electron. (China) **19**(4), 464–470 (1997)
32. X. Lv, G. Bi, C. Wan et al., Lv's distribution: principle, implementation, properties, and performance. IEEE Trans. Signal Process. **59**(8), 3576–3591 (2011)
33. D. Mendlovic, H.M. Ozaktas, Fractional fourier transforms and their optical implementation: I. J. Opti. Soc. Am. A Opt. Image Sci. **10**(10), 1875–1881 (1993)
34. H.M. Ozaktas, D. Mendlovic, Fractional fourier transforms and their optical implementation: II. J. Opti. Soc. Am. A Opt. Image Sci. **10**(12), 2522–2531 (1993)
35. H. Xu, Z. Yang, M. Tian et al., An extended moving target detection approach for high-resolution multichannel SAR-GMTI systems based on enhanced shadow-aided decision. IEEE Trans. Geosci. Remote Sens. **56**(2), 715–729 (2018)
36. R.L. Fante, Analysis of the displaced-phase-center radar for clutter reduction. Mitre Corporation Report, 1989, MT10666 (1989)
37. L. Lightstone, D. Faubert, G. Rempel, Multiple phase centre DPCA for airborne radar, in *IEEE National Radar Conference, Los Angeles* (1991), pp. 36–40
38. H. Wang, Mainlobe clutter cancellation by DPCA for space-based radars, in *Aerospace Applications Conference, Crested Butte* (1991), pp. 124–128
39. D. Cerutti-Maori, I. Sikaneta, A generalization of DPCA processing for multichannel SAR/GMTI radars. IEEE Trans. Geosci. Remote Sens. **51**(1), 560–572 (2013)

40. S. Li, Z. Li, Z. Liu, et al., Multichannel-two pulse cancellation method based on NLCS imaging for bistatic forward-looking SAR, in *IEEE International Geoscience and Remote Sensing Symposium, Yokohama* (2019), pp. 2229–2232

41. T.J. Nohara, Comparison of DPCA and STAP for space-based radar, in *IEEE International Radar Conference, Alexandria* (1995), pp. 113–119

42. R.S. Blum, W.L. Melvin, M.C. Wicks, An analysis of adaptive DPCA, in *IEEE National Radar Conference, Ann Arbor* (1996), pp. 303–308

43. Y. Chen, B. Qian, S. Wang, DPCA motion compensation technique based on multiple phase centers, in *IEEE CIE International Conference on Radar, Chengdu* (2011), pp. 711–714

44. L.E. Brennan, L.S. Reed, Theory of adaptive radar. IEEE Trans. Aerosp. Electron. Syst. **9**(2), 237–252 (1973)

45. I.S. Reed, J.D. Mallett, L.E. Brennan, Rapid convergence rate in adaptive arrays. IEEE Trans. Aerosp. Electron. Syst. **10**(6), 853–863 (1974)

46. J. Ward, Space-time adaptive processing for airborne radar, in *International Conference on Acoustics, Speech, and Signal Processing* (1995), pp. 2809–2812

47. R. Klemm, *Principles of space-time adaptive processing*, 3rd edn. (IEE Press, London, 2006)

48. J.H.G. Ender, Space-time adaptive processing for synthetic aperture radar, in *IEE Colloquium on Space-Time Adaptive Processing, London*, (1998), pp. 6/1–6/18

49. D. Cerutti-Maori, I. Sikaneta, C.H. Gierull, Optimum SAR/GMTI processing and its application to the radar satellite RADARSAT-2 for traffic monitoring. IEEE Trans. Geosci. Remote Sens. **50**(10), 3868–3881 (2012)

50. I. Sikaneta, C.H. Gierull, D. Cerutti-Maori, Optimum signal processing for multichannel SAR: with application to high-resolution wide-swath imaging. IEEE Trans. Geosci. Remote Sens. **52**(10), 6095–6109 (2014)

51. R. Klemm, Adaptive airborne MTI: an auxiliary channel approach. Commun. Radar Signal Process. IEE Proc. F **134**(3), 269–276 (1987)

52. Z. Bao, G. Liao, R. Wu et al., 2-D temporal-spatial adaptive clutter suppression for phase array airborne radars. Acta Electron. Sin. **21**(09), 1–7 (1993)

53. J. Ward, *Space-Time Adaptive Processing for Airborne Radar Systems* (MIT Lincoln Lab., Lexington, MA, 1994)

54. Y. Wang, Y. Peng, *Space-Time Adaptive Processing* (Tsinghua University Press, Beijing, 2000)

55. R. Klemm, *Principles of Space-Time Adaptive Processing*, 3rd edn. (IEE Press, London, 2006)

56. J. R. Guerci, *Space-Time Adaptive Processing for Radar*, 2nd edn. (Artech House, 2015)

57. R. Klemm, Comparison between monostatic and bistatic antenna configurations for STAP. IEEE Trans. Aerosp. Electron. Syst. **36**(2), 596–608 (2002)

58. B. Himed, J.H. Michels, Y. Zhang, Bistatic STAP performance analysis in radar applications, in *IEEE Radar Conference, Atlanta* (2001), pp. 198–203

59. H. Shnitkin, Joint STARS phased array radar antenna, in *Aerospace and Electronics Conference, Dayton* (1994), pp. 142–150

60. E.F. Stockburger, D.N. Held, Interferometric moving ground target imaging, in *IEEE International Radar Conference, Alexandria* (1995), pp. 438–443

61. J.N. Entzminger, C.A. Fowler, JointSTARS and GMTI: past, present and future. IEEE Trans. Aerosp. Electron. Syst. **35**(2), 748–761 (1999)

62. R. Hendrix, Aerospace system improvements enabled by Modern Phased Array Radar-2008, in *IEEE Radar Conference, Rome* (2008), pp. 1–6

63. R.L. Haupt, Y. Rahmat-Samii, Antenna array developments: a perspective on the past, present and future. IEEE Antennas Propag. Mag. **57**(1), 86–96 (2015)

64. J.H.G. Ender, P. Berens, A.R. Brenner, L. Rossing, U. Skupin, Multi-channel SAR/MTI system development at FGAN: from AER to PAMIR. *IEEE International Geoscience and Remote Sensing Symposium*, vol. 3 (2002), pp. 1697–1701

65. H. Wilden, A.R. Brenner, The SAR/GMTI airborne radar PAMIR: technology and performance, in *2010 IEEE MTT-S International Microwave Symposium* (2010), pp. 534–537

66. H. Wilden, O. Saalmann, B. Poppelreuter, K. Letsch, A. Brenner, Design aspects of the experimental SAR/MTI system PAMIR, in *European Radar Conference, 2005 EURAD 2005* (2005), pp. 25–28

67. D. Cerutti-Maori, J. Klare, A.R. Brenner et al., Wide-area traffic monitoring with the SAR/GMTI system PAMIR. IEEE Trans. Geosci. Remote Sens. **46**(10), 3019–3030 (2008)
68. S. Suchandt, H. Runge, H. Breit et al., Automatic Extraction of traffic flows using TerraSAR-X along-track interferometry. IEEE Trans. Geosci. Remote Sens. **48**(2), 807–819 (2010)
69. S. Suchandt, H. Runge, A. Kotenkov, et al., Extraction of traffic flows and surface current information using Terrasar-X Along-track interferometry data, in *2009 IEEE International Geoscience and Remote Sensing Symposium* (2009), pp. II-17-II-20
70. L. Chen, Z. Xi, Research on clutter depression based on DPCA in SAR. Electron. Meas. Technol. **29**(3), 138–139 (2006)
71. I.G. Cumming, F.H. Wong, *Digital Processing of Synthetic Aperture Radar Data: Algorithms and Implementation* (Artech House, Norwood, 2005)
72. Z. Li, S. Li, Z. Liu, H. Yang, J. Wu, J. Yang, Bistatic Forward-Looking SAR MP-DPCA Method for Space–Time Extension Clutter Suppression. IEEE Trans. Geosci. Remote Sens. **58**(9), 6565–6579 (2020)
73. D. Zhu, Y. Li, Z. Zhu, A keystone transform without interpolation for SAR ground moving-target imaging. IEEE Geosci. Remote Sens. Lett. **4**(1), 18–22 (2007)
74. R. Wang, O. Loffeld, Y. Neo, H. Nies, I. Walterscheid, T. Espeter, J. Klare, J. Ender, Focusing Bistatic SAR data in airborne/stationary configuration. IEEE Trans. Geosci. Remote Sens. **48**(1), 452–465 (2010)
75. J. Wu, Z. Sun, Z. Li, Y. Huang, J. Yang, Z. Liu, Focusing translational variant bistatic forward-looking SAR using keystone transform and extended nonlinear chirp scaling. Remote Sens. **8**(10), 840 (2016)
76. X. Zhang, G. Liao, S. Zhu, C. Zeng, Y. Shu, Geometry-information-aided efficient radial velocity estimation for moving target imaging and location based on radon transform. IEEE Trans. Geosci. Remote Sens. **53**(2), 1105–1117 (2015)
77. L. Chen, D. An, X. Huang, G. Zhu, A NLCS focusing approach for low frequency UWB one-stationary bistatic SAR, in *Proceedings of 16th International Radar Symposium (IRS)*, (Jun 2015) pp. 1082–1087
78. M.-J. Park, O.-M. Kwon, Stability and stabilization of discretetime T-S fuzzy systems with time-varying delay via Cauchy-Schwartzbased summation inequality. IEEE Trans. Fuzzy Syst. **25**(1), 128–140 (2017)
79. Y. Zhang, B. Himed, Effects of geometry on clutter characteristics of bistatic radars, in *Proceedings of IEEE Radar Conference, Huntsville, AL* (May 2003), pp. 417–424
80. B.L. McKinley, K.L. Bell, Range-dependence compensation for bistatic STAP using focusing matrices, in *Proceedings of IEEE Radar Conference (RadarCon)* (May 2015), pp. 1750–1755
81. J.N. Ash, An autofocus method for backprojection imagery in synthetic aperture radar. IEEE Geosci. Remote Sens. Lett. **9**(1), 104–108 (2011)
82. C.G. Broyden, A class of methods for solving nonlinear simultaneous equations. Math. Comput. **19**(92), 577–593 (1965)
83. M.D. González-Lima, F.M. de Oca, A Newton-like method for nonlinear system of equations. Numer. Algorithms **52**(3), 479 (2009)
84. J. Kennedy, R. Eberhart, Particle swarm optimization, in *International Conference on Neural Networks, Perth* (1995), pp. 1942–1948
85. M. Daneshyari, G.G. Yen, Cultural-based multiobjective particle swarm optimization. IEEE Trans. Syst. Man Cybern. Part B (Cybern.) **41**(2), 553–567 (April 2011)
86. W. Hu, G.G. Yen, Adaptive multiobjective particle swarm optimization based on parallel cell coordinate system. IEEE Trans. Evol. Comput. **19**(1), 1–18 (2015)
87. Z. Liu, Z. Li, H. Yu, J. Wu, Y. Huang, J. Yang, Bistatic forwardlooking sar moving target detection method based on joint clutter cancellation in echo-image domain with three receiving channels. Sensors **18**(11) (2018)
88. M.I. Skolink, *Radar Handbook*, 3rd edn. (McGraw-Hill, New York, 2008)
89. H. Meng, H. Zhang, G. Li, A class of novel STAP algorithms using sparse recovery technique (2009). http://arXiv.org/abs/0904.1313v1

90. Z. Yang, L. Xie, On gridless sparse methods for line spectral estimation from complete and incomplete data. IEEE Trans. Signal Process. **63**(12), 3139–3153 (2015)
91. Z. Liu, H. Yu, Z. Li, J. Wu, Y. Huang, J. Yang, Non-stationary clutter suppression approach based on cascading cancellation for bistatic forward-looking SAR, in *IEEE Radar Conference (RadarConf)* (2019), pp. 1–5
92. J.A. Tropp, A.C. Gilbert, M.J. Strauss, Algorithms for simultaneous sparse approximation. Part I: Greedy Pursuit. Signal Process. **86**(3), 572–588 (2006)
93. J.A. Tropp, Algorithms for simultaneous sparse approximation, part II: convex relaxation. Signal Process. **86**(3), 589–602 (2006)
94. K.Q. Duan, Z.T. Wang, W.C. Xie et al., Sparsity-based STAP algorithm with multiple measurement vectors via sparse Bayesian learning strategy for airborne radar. IET Signal Process. **11**(5), 544–553 (2017)
95. S.F. Cotter, B.D. Rao, K. Engan, K. Kreutz-Delgado, Sparse solutions to linear inverse problems with multiple measurement vectors. IEEE Trans. Signal Process. **53**(7), 2477–2488 (July 2005)
96. Yantai, Research on the experience model of ground clutter backscatter coefficient. Ship Electron. Eng. (2011)
97. Z. Liu, H. Ye, Z. Li, et al., Optimally matched space-time filtering technique for BFSAR nonstationary clutter suppression. IEEE Trans. Geosci. Remote Sens. **60**, 1–17, Art no. 5210617. (2022)
98. Z. Li, H. Ye, Z. Liu, et al., Bistatic SAR clutter-ridge matched STAP method for non-stationary clutter suppression. IEEE Trans. Geosci. Remote Sens. **60**, 1–14, Art no. 5216914. (2022)

Printed in the United States
by Baker & Taylor Publisher Services